THE EVOLUTION OF MATHEMATICS

RSA·STR

THE RSA SERIES IN TRANSDISCIPLINARY RHETORIC

Edited by
Michael Bernard-Donals *(University of Wisconsin)* and
Leah Ceccarelli *(University of Washington)*

Editorial Board:
Diane Davis, The University of Texas at Austin
Cara Finnegan, University of Illinois at Urbana-Champaign
Debra Hawhee, The Pennsylvania State University
John Lynch, University of Cincinnati
Steven Mailloux, Loyola Marymount University
Kendall Phillips, Syracuse University
Thomas Rickert, Purdue University

The RSA Series in Transdisciplinary Rhetoric is a collaboration with the Rhetoric Society of America to publish innovative and rigorously argued scholarship on the tremendous disciplinary breadth of rhetoric. Books in the series take a variety of approaches, including theoretical, historical, interpretive, critical, or ethnographic, and examine rhetorical action in a way that appeals, first, to scholars in communication studies and English or writing and, second, to at least one other discipline or subject area.

Other titles in this series:
Nathan Stormer, *Sign of Pathology: U.S. Medical Rhetoric on Abortion, 1800s–1960s*
Mark Longaker, *Rhetorical Style and Bourgeois Virtue: Capitalism and Civil Society in the British Enlightenment*
Robin E. Jensen, *Infertility: A Rhetorical History*
Steven Mailloux, *Rhetoric's Pragmatism: Essays in Rhetorical Hermeneutics*
M. Elizabeth Weiser, *Museum Rhetoric: Building Civic Identity in National Spaces*
Chris Mays, Nathaniel A. Rivers and Kellie Sharp-Hoskins, eds., *Kenneth Burke + The Posthuman*
Amy Koerber, *From Hysteria to Hormones: A Rhetorical History*
Elizabeth C. Britt, *Reimagining Advocacy: Rhetorical Education in the Legal Clinic*
Ian E. J. Hill, *Advocating Weapons, War, and Terrorism: Technological and Rhetorical Paradox*
Kelly Pender, *Being at Genetic Risk: Toward a Rhetoric of Care*
James L. Cherney, *Ableist Rhetoric*
Susan Wells, *Robert Burton's Rhetoric: An Anatomy of Early Modern Knowledge*
Ralph Cintron, *Democracy as Fetish*
Maggie M. Werner, *Stripped: Reading the Erotic Body*
Timothy Johnson, *Rhetoric, Inc: Ford's Filmmaking and the Rise of Corporatism*
James Wynn and G. Mitchell Reyes, eds., *Arguing with Numbers: Essays at the Intersections of Rhetoric and Mathematics*
Ashely Rose Mehlenbacher, *On Expertise: Cultivating Character, Goodwill, and Practical Wisdom*
Stuart J. Murray, *The Living from the Dead: Disaffirming Biopolitics*

G. Mitchell Reyes

THE EVOLUTION OF MATHEMATICS

A Rhetorical Approach

THE PENNSYLVANIA STATE UNIVERSITY PRESS
UNIVERSITY PARK, PENNSYLVANIA

Library of Congress Cataloging-in-Publication Data

Names: Reyes, G. Mitchell, author.
Title: The evolution of mathematics : a rhetorical approach /
   G. Mitchell Reyes.
Other titles: RSA series in transdisciplinary rhetoric.
Description: University Park, Pennsylvania : The Pennsylvania
   State University Press, [2022] | Series: The RSA series in
   transdisciplinary rhetoric | Includes bibliographical
   references and index.
Summary: "Applies contemporary rhetorical analysis to
   mathematical discourse, calling into question the commonly
   held view that math equals truth. Explores how mathematical
   innovation has historically relied on rhetorical practices of
   making meaning, such as analogy, metaphor, and
   invention"—Provided by publisher.
Identifiers: LCCN 2022026910 | ISBN 9780271094014
   (hardback) | ISBN 9780271094021 (paper)
Subjects: LCSH: Mathematics—Social aspects. | Rhetoric—
   Social aspects. | AMS: History and biography—History of
   mathematics and mathematicians—Miscellaneous topics.
Classification: LCC QA10.7 .R48 2022 | DDC 510—dc23/
   eng20220823
LC record available at https://lccn.loc.gov/2022026910

Copyright © 2022 G. Mitchell Reyes
All rights reserved
Printed in the United States of America
Published by The Pennsylvania State University Press,
University Park, PA 16802-1003

© The Rhetoric Society of America, 2022

The Pennsylvania State University Press is a member
of the Association of University Presses.

It is the policy of The Pennsylvania State University Press to
use acid-free paper. Publications on uncoated stock satisfy the
minimum requirements of American National Standard for
Information Sciences—Permanence of Paper for Printed
Library Material, ANSI Z39.48-1992.

For my parents, without whom I would not be me.
For my wife, Amy, without whom I would not be happy.
For my son, Lucca, without whom I would not be Papa.
And, finally, for all my fabulous students and all my wonderful teachers:
This one's for you.

Contents

Acknowledgments | ix

Introduction: Stranger Relations | 1

1 Plato's Mythos of Mathematics | 11

2 Imbrications: Mathematics as a Translative Rhetorical Force | 36

3 Transgressing the Limit: Invention, the Calculus, and the Rhetorical Force of the Infinitesimal | 60

4 How Imaginaries Became Real | 85

5 Algorithmic Culture and Economies of Translation | 103

6 Implications: Translative Rhetoric Revisited | 130

Notes | 151

Bibliography | 173

Index | 185

Acknowledgments

It was during an undergraduate rhetorical theory course that I, as a young mathematician, had my eyes opened to the possibility of thinking about math from a rhetorical perspective—to considering math for the first time as a language that persuades us to see and interact with the world in particular ways. For that class, which changed the course of my studies, my career, and my life, I am eternally grateful to Professor David Douglass, who remains to this day one of the best teachers to have ever graced the college classroom. Several years later, as a young graduate student, I laid down the first inklings of the ideas that would guide this book. I want to give special thanks to Professors Stephen H. Browne, Christopher L. Johnstone, J. Michael Hogan, Rosa Eberly, Thomas W. Benson, Charles E. Scott, and John Sallis for their care and patience with me and my ideas as they slowly emerged, taking the shape they eventually did only because of the endless hours these professors spent reading early drafts and engaging with me in long discussions. Each of you, in different ways, helped shape the professor I would eventually become, and for that I cannot thank you enough.

While you learn much from your professors in the university, you learn as much, if not more, from your peers, and so I must also thank several of my fellow graduate students with whom I was lucky enough to share classes, ideas, and even a few beverages. To Bradford Vivian, David Schulz, Pat Gehrke, Gina Ercolini, Christine Harold, Roger Stahl, Dave Tell, Dan Smith, and Michael Tumolo—thank you for keeping me sane, listening to my half-formed arguments, engaging with me in esoteric theory-babble, and modeling for me the practices of compassionate critical inquiry.

They say it takes a village to raise a book, and we can see that is true, but it also takes strong academic institutions, which support the slow, laborious process of academic scholarship. For that support I must recognize my own institution, Lewis and Clark College, for the resources and sabbatical time given, without which this book could not have been written. Through my institution and the many others I have visited around the globe, I have been fortunate to connect with amazing colleagues who have helped advance many of the ideas in this book. I would like to offer special thanks to Kendall Phillips, Nathaniel Cordova, Craig

Smith, David Schulz, Bradford Vivian, Kundai Chirindo, Heather Hayes, Joshua Hanan, Catherine Chaput, Eva Reyes, James Wynn, Jason Barrett-Fox, Rob Topinka, Paul Allen, and Iva Stavrov for the enumerable conversations, email exchanges, and reading groups, all of which helped me inch along, however slowly, toward the finish line that is this book. It is my great good fortune to call each of you colleagues.

Of course, none of this collegial and institutional support would amount to much without my family, a network of love and understanding I try hard not to take for granted. To my parents, Wally Newberry and Marta Reyes-Newberry, what can I say? Thank you for showing me the way. You mean the world to me. To my brothers and sisters: you have each taught me more than you know. Thanks for your patience. To my nephews and nieces: each of you light up the room in so many ways. To my wife, Amy Wing, words fail me. For all those times you helped create space for the writing of this book, the rewriting of this book, the editing of this book, I can only say from the deepest parts of my heart—thank you.

Finally, I would like to thank the editors of this series, Leah Ceccarelli and Michael Bernard-Donals, and the two blind reviewers for their support and sage advice, as well as the editorial staff at the Pennsylvania State University Press and especially acquisitions editor Ryan Peterson, who has helped shepherd this book through the submission and revision process with kindness and enthusiasm.

Introduction | Stranger Relations

In 2010 a "Republican Wave" enabled the party to take Congress back from Democrats. Perhaps more important, it enabled Republicans, who also won widely at the state level, to redraw 211 congressional districts, which in the subsequent elections of 2012, 2014, and 2016 enabled the party to outperform its total vote share by 4–5 percent. The art of redrawing congressional districts is widely known as gerrymandering, and, while it is practiced by both major parties in US politics, over the past three decades Republican gerrymandering tactics have outperformed Democratic efforts. How do these districts end up looking like an "Upside-down Chinese Dragon" or a "Rabbit on a Skateboard"?[1] In short, operatives use sophisticated algorithms and statistical analyses to generate congressional maps advantageous to a particular party, the results of which have systematically increased polarization and political extremism in the United States.[2]

In January 2020 Robert Julian-Borchak Williams was arrested at his home in front of his wife and two young daughters. The arresting officers would not divulge why he was arrested, showing him a felony warrant for larceny instead. After being booked and spending a night in jail, Williams was shown the evidence "against" him, the primary piece of which was a surveillance photograph of a heavyset Black person the police suspected of shoplifting. When the police asked Williams if he was the person in the photo (which bore little resemblance to Williams), he said, "No, this is not me. You think all black men look alike?"[3] What Williams did not know at the time was that facial-recognition technology the Detroit bureau used had produced a flawed match, leading to his arrest. While police bureaus have used facial-recognition technology for decades, few in those bureaus understand that facial-recognition algorithms are notoriously poor at matching nonwhite faces.[4] And while the Detroit bureau did eventually release and apologize to Williams, they cannot erase his daughters' traumatic memory of their father in handcuffs on the front lawn.

In November 2018 gene editing in humans became real. He Jiankui, a Chinese scientist at Southern University of Science and Technology in Shenzhen, admitted to using CRISPR-CAS9 technology to edit the DNA of human embryos, which were then implanted in two women. His motives seemed pure—even benevolent: he simply sought to buffer the embryos against HIV so that these women could safely have children. His actions were nevertheless condemned by the global scientific community as premature and irresponsible, and he has since been fined and sentenced to three years in prison.[5] If anything, however, his actions appear to have accelerated genome editing trials in humans, with promising new approaches emerging for everything from cancer to blindness. Genomic editing of nonhuman entities has been even more prolific, as such study offers real potential for not just fighting diseases like malaria but eliminating them completely.[6] Doing so, of course, entails a fundamental transformation of the global ecology (the elimination of malaria-carrying mosquitos), the ramifications of which scientists admit they do not understand and cannot predict.

What do these three seemingly disconnected events have in common? Each of them, in different ways, bespeak the rising influence of mathematical discourse on everything from democracy to surveillance and racial bias to ecological formation. In gerrymandering one finds the mathematics of geometric optimization used to affect massive shifts in political power, in facial-recognition technology one finds sophisticated learning algorithms that nevertheless spread the built-in biases of their creators, and in CRISPR applications one finds a powerful application of the mathematical apprehension of "life" in something humans now casually refer to as DNA, which may author a whole new algebra of creative evolution that is both terrifying and, potentially, beautiful. At the same time, these events also bespeak the strangeness of our world—the radical difference and differencing of life (not just human life but *all* life) that is happening at such a rapid rate we can actually observe it within the geologically miniscule confines of a human life span.

Some like to think that below the surface of all this apparent change runs a river of essential sameness, that the challenges of today are not that different from the challenges of the past. In gerrymandering one sees the same old political games; in facial-recognition technology, the same old racial biases. And, of course, in a way those people are absolutely right. Yet I find that line of thinking to be a dangerous half-truth, for it masks a larger phenomenon, one not at the level of the individual or even the collective. No, the phenomenon I have in mind is happening at the level of the vincula—the fundamental relations that compose our social-material world.[7] The strangeness (the alterity in Jacques Derrida's sense) of our world lies there—not in our individuated experience but in the decomposition and recomposition of basic social-material relations that were so invisibly stable that previous generations took them largely for granted. We are talking here about long-enduring differentiating relations between the organic and the mechanic, the animate and the inanimate, the actual and the virtual, the sentient and the nonsentient, the human and the nonhuman being blurred, altered, and upended.[8] We're talking about a world in which humans have become so influential in their collective behavior that they are starting to bend and distort the curvature of biological evolution.[9] We're talking about a world in which the biological and the mechanical are so richly intertwined that novel speciation becomes possible, that a merger of human intelligence and artificial intelligence is not only probable but perhaps even necessary. We're talking about a world in which we can weave the real and the virtual into such beautiful and sublime networks that the distinction becomes not only arbitrary but, more important, an obstacle to understanding our new proliferating realities.

How can we come to an empowering understanding of this new strange world in which we find ourselves? My hope is that this book provides one avenue toward such an understanding. That understanding begins with close rhetorical analysis of the history and evolution of mathematical discourse. There, in the archives of mathematical practice, we find many of the ingredients of our contemporary age. And, most crucially, we will find them not in their final polished form but in their first moments of emergence, before they have been hardened and refined into seemingly transcendent kernels of truth—or, as some have labeled them, black boxes.[10]

In the study of mathematical discourse and its many genealogies, we also see that mathematics does not do what we are so often told it does. We are told that, at its best, mathematics reveals absolute truth—what the Greeks called *aletheia*— timeless patterns and forms governing an ordered cosmos. Thus the renowned Isaac Asimov wrote, "Now we can see what makes mathematics unique. Only in mathematics is there no significant correction—only extension. Once the Greeks had developed the deductive method, they were correct in what they did, correct for all time. Euclid was incomplete and his work has been extended enormously, but it has not had to be corrected. His theorems are, every one of them, valid to this day."[11] This is the mathematics of timeless transcendental truth (*pure mathematics*), which continues to hold sway both within and outside of contemporary mathematical communities (a point more fully developed in chapters 1 and 2). Study of mathematical discourse and practice, however, reveals a much different enterprise. Instead of a practice of disembodied reason purifying human thought and, in the process, eliminating the all-too-human rhetorical components of subjective judgment, desire, ideology, and history, one finds a discursively governed practice of embodied symbolic inscription in which myriad agencies (human, discursive, nonhuman) occasionally congeal into principles of composition with the capacity to transform the vincula that compose our world.[12] Mathematics, in this sense, is far more interesting than a truth-production machine. Instead, it is perhaps the most powerful means we humans have so far conjured for symbolically transforming the social-material world, which means if we hope to understand the world we are helping write ourselves into, we must come to understand the translative rhetorical force of mathematical discourse.

What does it mean to describe math as a translative rhetorical force? First, it does not mean mathematics is secretly just another form of persuasion or manipulation. Surely mathematics can be put to such ends, as many scholars have shown, but in describing mathematics as a translative rhetorical force I have something different in mind.[13] The senses of rhetoric as oratory or persuasion or, in Aristotle's words, "the faculty of observing in any given case the available means of persuasion" must be temporarily put aside if we are to understand the translative force of mathematical discourse.[14] More contemporary theories of rhetoric as argument or symbolic action get us closer but remain too committed to theories of social constructivism, which tend to divide nature from culture.[15] This is not to say we must reject these senses of rhetoric but instead acknowledge that rhetoric is not and has never been a singular entity. Rhetoric is not "big"; it is diverse, and each rhetorical ecology has its own unique discursive practices, capacities of address, and means of enabling and constraining traffic between symbolicity and materiality.[16] And it is ultimately that traffic that we are after in this book.

To describe math as a translative rhetorical force, then, suggests a certain kind of shift in our thinking about both rhetoric and mathematics. Associating mathematics primarily with the study of timeless, ahistorical, a priori objects (as mathematical realists do) has alienated it from rhetoric, wherein history, culture, and context are crucial and must be taken seriously.[17] Likewise, associating rhetoric primarily with persuasion has long alienated it from the mathematical arts of demonstrative proof and refutation. None of these associations, however, are permanent or essential, and one might say the liveliness of both rhetoric and math (as fields of study) can be measured by their capacities for evolution and diversification. The case I make throughout this book is that if we can rethink both in terms of the vinculum—the relations that form the networks that compose the social-material world—we can escape their estrangement and begin to take their increasingly consequential interations seriously.

Thinking of both rhetoric and mathematics through the lens of relationality has real potential. Not only can it offer us an avenue of escape from the pieties of mathematical realism, but, more important, it promises a better explanation and understanding of both how mathematics evolves and how it has become the metadiscourse of the twenty-first century. Mathematical realism, as many scholars have noted, fails to explain the growth of mathematics. In Imre Lakatos's widely regarded book *Proofs and Refutations*, for example, he notes that formalism (one contemporary version of mathematical realism) cannot explain math's evolution because it "denies the status of mathematics to most of what has been commonly understood to be mathematics, and can say nothing about its growth. None of the 'creative' periods and hardly any of the 'critical' periods of mathematical theories would be admitted into the formalist heaven, where mathematical theories dwell like the seraphim, purged of all the impurities of earthly uncertainty."[18] The ideology of purification in mathematics—which can be found in most arguments for mathematical infallibility—comes at a high cost, blinding us from the very practices that enable the emergence and articulation of new mathematical relations (like $i = \sqrt{-1}$) with enough agential force to unleash not only whole new realms of math but also whole new worlds of social-material relations. Without rhetorical study of mathematical practice and mathematical discourse, these genealogies would remain shrouded in the mythos of mathematical realism to which Asimov gave voice, from whence the power of mathematics is both apotheosized and poorly understood. The problem is that today, in an environment profoundly shaped by mathematical discourse, that ignorance has increasingly dire consequences.

From the rhetorical perspective offered here, however, mathematics becomes both better understood and more interesting. Instead of a seemingly dead language of absolute truth produced by the rare genius with the appropriate combination of imagination, deductive logic, and pure reason, mathematics transforms from a static collection of discovered tools (or abstract truths) into a dynamic practice of creative innovation. The rhetorical approach to math developed here encourages one to see mathematics as a collection of powerful translation machines, or a collection of languages and ways of thinking (for every language is a way of thinking) that translate those who learn to "speak" them just as much as they do the world that such "speaking" rearticulates. From this vantage point, one might ask not just "What problems can mathematics solve?" but also "Who do we become when we think mathematically?"; "What becomes of the world through a particular mathematical lens?"; and "How do particular mathematical networks function as principles of composition?"

The good news is that some mathematicians have intuitively understood this rhetorical approach and practiced it both in their doing of math and in their pedagogy. Philip Davis and Reuben Hersh, who first challenged the dissociation between rhetoric and mathematics in the 1980s, were practicing mathematicians.[19] Lakatos's classic *Proofs and Refutations* remains one of the best to capture the dynamic role of analogy, metaphor, historical context, and argument in the practice of thinking mathematically. And as the great mathematician William Thurston observed in 1994, "If what we are doing is constructing better ways of thinking, then psychological and social dimensions are essential to a good model for mathematical progress."[20] Thurston went on in that essay to illustrate the point with the concept "derivative," for which there are at least seven different meanings. The meaningfulness of "derivative," according to Thurston, comes not from each particular definition but from its unlimited polysemy. Mathematicians, Thurston argued, perhaps misunderstand how mathematics progresses when they call for one single, consistent definition for mathematical concepts. In the history of mathematics too one finds that rhetoric and math were once closely linked and thought to be mutually beneficial.[21] One final bit of good news: the hegemony of mathematical realism is beginning to break down, and there are many movements afoot within the mathematics community that advocate for one aspect or another of this rhetorical paradigm, whether one considers the emphasis ethnomathematics puts on culture and place or the emphasis some constructivist mathematicians put on practices of signification such as metaphor and analogy.[22]

Likewise in the field of rhetorical studies, many scholars have become increasingly interested in how mathematical discourse shapes public culture, and that research has converged with increasing interest in the field in the materiality of rhetorical practice (what some refer to as "new materialism").[23] George Kennedy opened the way for a more materialist-oriented theory of rhetoric with his essay "A Hoot in the Dark," which offered a radically different conception of rhetoric as the study of "rhetorical energy" rather than persuasion or argument or symbolic action. Words, Kennedy explains, "involve different degrees of expenditure of physical energy in their utterance. Rhetoric in the most general sense may perhaps be identified with the energy inherent in communication: the emotional energy that impels the speaker to speak, the physical energy expended in the utterance, the energy level coded in the message, and the energy experienced by the recipient in decoding the message."[24] Just a year later Dilip Gaonkar published his now famous critique (in rhetorical studies) of the widespread impulse to treat rhetoric as a metatheoretical discourse that can be applied to all forms of symbolic action.[25] While many interpreted Gaonkar's critique (which focused on the then rhetoric of science literature) as a rejection of the broad expansion of rhetorical studies beyond the examination of public discourse, his real target was scholarly neglect of the diversity of rhetoric. To say as much is to say nothing new. Alan Gross, for example, described Gaonkar's essay as a "wonderful destructive critique of rhetoric of science" whose main argument was that "a technology, namely classical rhetoric, that was designed to teach young boys on how to give speeches in public forums, was not going to be a technology that could be easily transferred into a critical methodology, especially one of science, which was a whole lot more complex."[26] Rhetoric (symbolic action) is not just big, in other words; it is diverse, and rhetorical scholars need to develop critical tools capable of addressing the diversity and specificity of different discursive realms. Put differently, if one treats the discourses of math and science with the same theories and concepts developed to produce and analyze public discourse, one will simultaneously "universalize rhetoric" as a metadiscourse and, as a result, carry into one's work a "trained incapacity" to see the substantive differences between math, science, and oratory as altern modalities of discursive action. The result is likely to be heavy on imposition of critical theories and concepts that are ill-equipped for understanding the practices of math and science, not to mention their agential force in the world.[27] To do justice to the full complexity of rhetoric, Gaonkar argued, we must resist the humanist "impulse to universalize rhetoric," which historically has placed the human at its origin.[28]

While these essays were initially either rejected or critically marginalized, they nevertheless created space to challenge deeply anthropocentric theories of rhetoric and begin developing alternatives. They did this not by challenging the field on epistemological or ideological grounds but by challenging the ontological hegemony of "speech" within rhetorical studies. For if one orients rhetoric's ontology around speech (traditionally conceived), then rhetoric's existence (its being) becomes inextricably tied to humans and their "faculties" of speech, which of course necessarily excludes all nonhumans from the proper domain of rhetoric (we can see this bifurcation codified, in fact, in Aristotle's distinction between "artistic" and "inartistic" proofs).[29] Fortunately, an increasing core of leading scholars have worked to diversify rhetoric's ontology not as a means of rejecting speech but as a means to address the diversity and complexity of rhetoric in the twenty-first century.[30] Challenges to the field's anthropocentrism and logocentrism have come in many forms, but a great majority, as Diane Davis and Michelle Ballif observe, invite "our attention to the various ecologies that instantiate any so-called rhetorical situation, including material geologies as well as networked relations."[31] And as scholars progressively decentered the human in our theories, whole fields of rhetorical action previously obscured slowly came into view. Suddenly, we could start to think and write seriously about how animals communicate, how science rhetorically weaves humans and nonhumans into increasingly novel hybrids, and how networked ecologies converse, persuade, and share in surprisingly complex ways.[32]

These broad shifts in rhetorical studies are crucial to the study of mathematical discourse in at least three ways: first, to understand mathematical discourse and practice, one must first decenter human agency. As most who have studied and practiced mathematics know, when one "does" mathematics one is not alone, and—equally important—one is not free to do whatever one wishes.[33] This in part is what has kept rhetoric and mathematics estranged. Rhetorical scholars often operate—whether consciously or unconsciously—as if humans have a monopoly on rhetorical agency, as if our freedom and our ontological significance as humans derives from our capacity to speak, persuade, and therefore *act* instead of simply *move*.[34] Yet the practice of doing mathematics reveals rather quickly to most (not just mathematicians) how little an anthropocentric theory of symbolic action applies to mathematical practice. In mathematics one is not free to do whatever one wants (whether claiming $2 + 2 = 5$ or that triangles have 190 degrees). Instead we *think with and through* various mathematical symbolics (number, geometry, trigonometry), collaborating with an altern agency, which often shapes and refines our thinking and, occasionally, surprises us right out of our seats.[35] Immediately apparent in the doing of math, then, is a kind of *distributed agency* that is at odds with anthropocentric theories of rhetoric.

Second, the realization that agency is distributed in the practice of mathematics encourages one to consider the agential force of mathematical discourse itself—an agential force excessive to the agency or intentions of any particular mathematician. When we genealogically trace the emergence of the Calculus (chapter 3) or imaginary numbers (chapter 4), in other words, we begin to see how mathematics has evolved in part by going beyond the logics of representation that dominated mathematics for millennia.[36] Those logics, perhaps best captured in Euclid's

Elements, both enabled and constrained mathematical thought, and it was only their eclipse that has brought us to our current techno-scientific age. In these studies of the Calculus or imaginary numbers, posthumanist theories of discourse as constitutive are crucial for unpacking the diffractive capacities of particular mathematical statements, which—once unpacked—reveal themselves as complex networks of relations with their own judgment parameters and ontological force.[37]

Finally, the notion of agency as distributed opens the way to thinking not only about the various agencies at play in mathematical discourse but also how they congeal into mathematical phenomena with sufficient agency to function as principles of composition, introducing novel relations into our social-material world and thereby accelerating the becoming (the biocultural evolution) of that world. This point is crucial if one hopes to shed light not just on the symbolic action constitutive of technical mathematical statements but also on math's increasing translative force in the twenty-first century. That translative force emerges not from discovery of "the real" but from the translation, reconfiguration, and occasional expansion of the real. Mathematics' translative force, in other words, emerges not from the blinding light of timeless truth but from the study and translation of worldly relations into various commensurable symbolic forms, which then enables terrifically inventive practices of metaphorical and analogical concept-stretching that accelerate the production of new math and novel relational configurations. And it is those novel relational configurations that increasingly shape everything from public cultures (gerrymandering, racialization) to global ecologies (genomic editing).

Rhetorical study of mathematics has, up to this point, been primarily interested in the ways mathematics influences public discourse and public culture—what Edward Schiappa describes as the "rhetoric *of* mathematics."[38] This can include everything from "the impact of math on policy making" to "interests in how mathematization reshapes technical fields like Biology or Meteorology."[39] To a lesser extent rhetorical scholars have also investigated the "rhetoric *in* mathematics," which focuses on the modes of argument and persuasion in mathematical practice.[40] These studies have illuminated many aspects of mathematical invention and influence, yet they are also limited (my own work included) by their implicit or explicit understanding of rhetoric as the study of symbolic action, whether in the form of speech, argument, persuasion, analogy, or metaphor. In proposing that we think of rhetoric through the concept of the vinculum, I am seeking a way beyond the episteme of representation; I am seeking a language of understanding for the traffic between symbolicity and materiality that mathematical discourse and practice enable. Understanding that movement, I argue, will not only advance rhetorical theory; it will ultimately enable rhetorical scholars to make good on the charge of providing critical insight into the formation, circulation, and transformation of public culture, which we increasingly see is not a product of symbolic action alone but an interweaving of symbolic-material relations into increasingly complex networks, the totality of which we call reality.

The path toward understanding the translative rhetorical force of mathematical discourse begins with their estrangement. That estrangement did not commence with Plato (the Pythagoreans influenced him heavily after all), but his dialogues did codify and harden their alienation, and so we begin with Plato's dialogues in order to see how rhetoric and mathematics were dissociated and what becomes of mathematics in Plato's metaphysics. Chapter 2 then traces various challenges to Platonic realism in the twentieth and twenty-first centuries as a means of developing the notion of *translative rhetoric*, a conception of rhetoric particularly useful for the study of mathematical discourse and practice. The two chapters that follow are broad genealogies of the Calculus (chapter 3) and imaginary numbers (chapter 4), from which we begin to see the evolution of mathematical discourse beyond the Euclidean enmeshed episteme of representation as well as how those transgressions connect with math's increasing translative rhetorical force. In chapter 5 we see the culmination of that translative force in a study of algorithmic culture and, specifically, the role algorithms played in the 2008 financial crisis. The book ends with discussion of how our new conceptions of rhetoric and mathematics help us understand an era many scientists have labeled the Anthropocene—a time of immense crisis and unparalleled possibility.

1

Plato's Mythos of Mathematics

Now that the study of calculation has been mentioned, I recognize how subtle it is and how in many ways it is useful to us for what we want. . . . It leads the soul powerfully upward and compels it to discuss numbers themselves. It won't at all permit anyone to propose for discussion numbers that are attached to visible or tangible bodies.
—Plato, *Republic* 525d

Plato cannot be avoided. One would simply be remiss to write a book about rhetoric and mathematics without engaging his work. And while countless books and articles have been written about Plato's philosophy of mathematics and his views of rhetoric, it is testament to his influence that few scholars have treated these topics in tandem. For it is out of Plato's considerable opus that mathematics and rhetoric are situated in radically different realms—mathematics in the realm of *episteme* (knowledge) and rhetoric in the realm of *doxa* (opinion). To upend their long estrangement, then, one must begin with Plato and the mythos of mathematics that emerges throughout his dialogues.

The epigraph offers a glimpse of how Plato weaved mathematics into his metaphysics. Already one can see intimate links between number, transcendence, and the soul. For Plato, mathematics in its truest form was never about practical application, something the last sentence implies and that he says explicitly a few lines later. Practical uses of numbers and calculation might be advantageous in some particular way, but Plato taught his readers not to get distracted by practicalities and lose sight of the pursuit of *aletheia* (truth). Rather than a mere tool, mathematics for Plato (especially geometry) was novel, cutting-edge, irrefutable evidence of his theory of forms (*Republic*; *Theaetetus*; *Timaeus*), the existence of the everlasting soul (*Meno*; *Phaedo*), and justification for the imposition of a geometric model of decision-making on the polis (*Protagoras*; *Gorgias*). For Plato, as for all Platonic realists to follow, there exists beneath the appearance of things (the perceptibles) an underlying and eternal mathematical reality. This realm of mathematical objects occupies the aporia between the everyday realm of perceptibles and the imperceptible realm of ideas. Mathematics is a powerful means by which one can *approach* the realm of ideal forms and provoke the soul to remember the true nature of things, which is why Plato engraved on the door of his academy, "Mèdeis ageômetrètos eisitô mou tèn stegèn [Let no one ignorant of geometry come under my roof]."[1]

Overstating Plato's influence on historical and contemporary views of mathematics would be difficult. Carl Boyer and Uta Merzbach note in their classic *A History of Mathematics* that "the Platonic Academy in Athens became the mathematical center of the world, and it was from this school that the leading teachers and research workers came during the middle of the fourth century B.C."[2] Over the roughly two and a half millennia since Plato's time, his ideas about mathematics have enjoyed incredible longevity. Ian Hacking wrote of Plato's continued influence among mathematicians, noting the embodiment of Platonic realism in the debate between Alain Connes (a prominent mathematician who received the Fields Medal in 1982 and the Clay Research Award in 2000) and Jean-Pierre Changeux (an eminent neurobiologist).[3] In that debate Connes expressed a thoroughly Platonic view of math, claiming at one point that "the working mathematician can be likened to an explorer who discovers the world."[4] Many of the best mathematicians and scientists in history likewise saw through Platonic eyes: Leopold Kronecker famously opined, "God made the integers and all else is the work of man," and Kurt Gödel, whose famous incompleteness theorems proved the fallibility of any nontrivial axiomatic system, formulated and defended Platonic realism throughout his career.[5] Even Albert Einstein, whose work ultimately challenged the Newtonian mathematical-physical paradigm, stated in a lecture to the Prussian Academy of Sciences that "one reason why mathematics enjoys special esteem, above all other sciences, is that its laws are absolutely certain and indisputable."[6] As mathematician-cum-semiotician Brian Rotman observes, "When they pursue their business, mathematicians do so neither as formalist manipulators nor as solitary mental constructors, but rather as scientific investigators engaged in publicly discovering objective truths."[7]

If we hope to understand and effectively challenge the alienation of rhetoric and mathematics, then, we must trace their progressive estrangement in Plato's dialogues. There we will see how Plato's thinking about mathematics evolves and how that evolution increasingly conceals the fascinating discursive materiality of mathematics in a dense mythos. The use of the concept "mythos" here should not be confused with the classical notion of mythos connected with the mythopoeic mindsets of ancient Greece. Instead, this chapter traces in Plato's dialogues the rhetorical construction of a mathematical mythos—a metaphysical construct of mathematics that subordinates the actual embodied practice of doing mathematics to a mythical means of approaching the ideal forms to which mathematics at its best (according to Plato) gestures. To extract these elements from Plato's dialogues, however, requires a resistive reading strategy—one that seeks to reveal precisely what an author (whether intentionally or unintentionally) conceals. We must, in other words, attempt to read Plato against Plato. In so doing we both come to a clearer understanding of the consequences of Plato's mythos of math and begin to see productive avenues of escape.

The Emergence of Platonic Realism

Though not the most sophisticated discussion of mathematics that Plato offers in his dialogues, Socrates's encounter with an enslaved Boy in *Meno* is, as David H. Fowler notes, "our *first* direct, explicit, extended piece of evidence about Greek mathematics."[8] As such, it offers a promising point of departure. Socrates wants to prove his claim to Meno that "the soul is immortal" (*Meno* 81c) and that discovery of truth is, properly understood, a form of recollection. He asks Meno to hail an unlearned slave to demonstrate (82b–d):

> SOCRATES: Tell me, boy, do you know that a square figure is like this?
> BOY: I do.
> SOCRATES: Now, a square figure has these lines, four in number, all equal?
> BOY: Certainly.
> SOCRATES: And these, drawn through the middle, are equal too, are they not?
> BOY: Yes.
> SOCRATES: And a figure of this sort may be larger or smaller?
> BOY: To be sure.
> SOCRATES: Now if this side were two feet and that also two, how many feet would the whole be? Or let me put it thus: if one way it were two feet, and only one foot the other, of course the space would be two feet taken once?
> BOY: Yes.
> SOCRATES: But as it is two feet also on that side, it must be twice two feet?

BOY: It is.

SOCRATES: Then the space is twice two feet?

BOY: Yes.

SOCRATES: Well, how many are twice two feet? Count and tell me.

BOY: Four, Socrates.

I cite the dialogue at length here because so much is revealed, both about Plato's philosophy of mathematics and mathematics as practiced in fourth century BCE Athens. Immediately one sees an intimate connection between Greek geometry and the material diagram Socrates draws for the Boy. That diagram is of two parts: the equal lines that form the square and the two perpendicular lines that divide it into four equal parts (see fig. 1). Note too the conspicuous absence of numbers or letters assigned to the diagram. We are a long way from the Cartesian notion of arithmetized geometric space that we take for granted today. Instead of proving the relation between two spaces through the tools of algebraic geometry ($a^2 + b^2 = c^2$), one must, to think *with* Socrates, prove these relations spatially—that is, demonstrate them geometrically with reference only to the diagram, a form purified of discursive content.[9] Perhaps that is why Plato later wrote in the *Republic* that geometers'"language is most ludicrous, though they cannot help it, for they speak as if they were doing something and as if all their words were directed towards action. For all their talk is of squaring and applying and adding and the like, whereas in fact the real object of the entire study is pure knowledge" (Plato, *Republic* 7.527a).

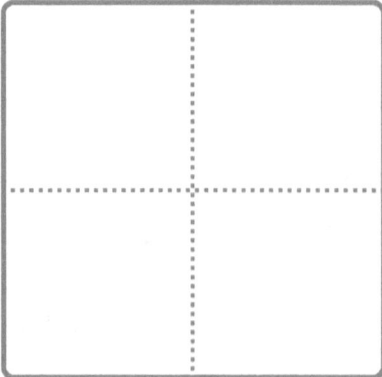

Fig. 1 | Bisected square

It is for the discovery of "pure knowledge" that Plato thought so highly of geometry—a nonarithmetic geometry that, through diagrams, enables demonstrations so compelling, so purified of desire or intention, that, as Aristotle put it, the solutions offered "could not be otherwise."[10] That is also why Socrates takes pains to restate the question he puts to the boy in spatial rather than arithmetic terms: "Or let me put it thus: if one way it were two feet, and only one foot the other, of course the space would be two feet taken once." How strange this sounds to the modern ear, but it captures a significant difference: for Plato and the Greeks of his time geometry was about relations between spaces and forms, not relations between the numbers (units, areas, etc.) those spaces might represent.[11]

The scene between Socrates and the Boy reveals even more as it develops (*Meno* 82d–e):

> SOCRATES: And might there not be another figure twice the size of this, but of the same sort, with all its sides equal like this one?
> BOY: Yes.
> SOCRATES: Then how many feet will it be?
> BOY: Eight.
> SOCRATES: Come now, try and tell me how long will each side of that figure be. This one is two feet long: what will be the side of the other, which is double in size?
> BOY: Clearly, Socrates, double.

Here Socrates has led the Boy into something of a trap. The modern student would simply solve for $\sqrt{8}$, but that again takes advantage of mathematical systems (arithmetic space, irrational numbers, algebra) not available to ancient Greeks. If we think spatially, as the Greeks did, it would be reasonable to believe that doubling the size of a square would require doubling the length of its sides, for the distinction between doubling and squaring would not be immediately apparent. And this sets up Socrates's demonstration of knowledge as a form of recollection (82e):

> SOCRATES: Do you observe, Meno, that I am not teaching the boy anything, but merely asking him each time? And now he supposes that he knows about the line required to make a figure of eight square feet; or do you not think he does?

> MENO: I do.
>
> SOCRATES: Well, does he know?
>
> MENO: Certainly not.
>
> SOCRATES: He just supposes it, from the double size required?
>
> MENO: Yes.
>
> SOCRATES: Now watch his progress in recollecting, by the proper use of memory.

In what follows, Socrates uses the diagram to show the Boy that doubling the length of line "fourfolds" the size of the original square. He then asks the Boy (83c–e):

> SOCRATES: What line will give us a space of eight feet? This one gives us a fourfold space, does it not?
>
> BOY: It does.
>
> SOCRATES: And a space of four feet is made from this line of half the length?
>
> BOY: Yes.
>
> SOCRATES: Very well; and is not a space of eight feet double the size of this one, and half the size of this other?
>
> BOY: Yes.
>
> SOCRATES: Will it not be made from a line longer than the one of these, and shorter than the other?
>
> BOY: I think so.
>
> SOCRATES: Excellent: always answer just what you think. Now tell me, did we not draw this line two feet, and that four?
>
> BOY: Yes.
>
> SOCRATES: Then the line on the side of the eight-foot figure should be more than this of two feet, and less than the other of four?
>
> BOY: It should.
>
> SOCRATES: Try and tell me how much you would say it is.
>
> BOY: Three feet.

One of the most fascinating aspects of Plato's dialogues is the way one can piece together the practices of ancient Greek thought even when the thrust of Plato's argument ultimately seeks to conceal those practices. In these lines one can detect a mathematical heuristic beginning to emerge, namely, the method of exhaustion, which is simply a sophisticated way of describing reasoned guess and check: the

Boy's original guess is tested and found wanting. But that guess, though wrong, is not a failure, for it provides a valuable upper limit that, combined with the original square as lower limit, creates a range of possibilities. The Boy then uses those upper and lower limits to make a more informed guess (incidentally, Archimedes famously used the method of exhaustion to approximate π to such a degree of accuracy (22/7ths) that π became known as the "Archimedes constant").[12]

While this method has its merits for approximating the unknown, it cannot discover the kind of pure knowledge Plato seeks. The method of exhaustion, for Plato, was mere calculation, and, while pragmatically useful, it paled in comparison to the pure, irrefutable knowledge that a geometric logos provides.[13] So Socrates once again returns to the diagram, demonstrating that a three-foot square is "thrice three feet" (*Meno* 83e) and so still off the mark. Socrates then asks (83e–84c),

> SOCRATES: But from what line shall we get it? Try and tell us exactly; and if you would rather not reckon it out, just show what line it is.
>
> BOY: Well, on my word, Socrates, I for one do not know.
>
> SOCRATES: There now, Meno, do you observe what progress he has already made in his recollection? At first he did not know what is the line that forms the figure of eight feet, and he does not know even now: but at any rate he thought he knew then, and confidently answered as though he knew, and was aware of no difficulty; whereas now he feels the difficulty he is in, and besides not knowing does not think he knows.
>
> MENO: That is true.
>
> SOCRATES: And is he not better off in respect of the matter which he did not know?
>
> MENO: I think that too is so.
>
> SOCRATES: Now, by causing him to doubt and giving him the torpedo's shock, have we done him any harm?
>
> MENO: I think not.
>
> SOCRATES: And we have certainly given him some assistance, it would seem, towards finding out the truth of the matter: for now he will push on in the search gladly, as lacking knowledge; whereas then he would have been only too ready to suppose he was right in saying, before any number of people any number of times, that the double space must have a line of double the length for its side.

MENO: It seems so.

SOCRATES: Now do you imagine he would have attempted to inquire or learn what he thought he knew, when he did not know it, until he had been reduced to the perplexity of realizing that he did not know, and had felt a craving to know?

MENO: I think not, Socrates.

SOCRATES: Then the torpedo's shock was of advantage to him?

MENO: I think so.

For the first time in this scene one begins to understand the motives behind Plato's use of geometry—not merely to demonstrate a geometric proof but to justify his broader metaphysical views on dialectic, the immortality of the soul, knowledge as recollection, and ultimately his theory of Forms (which are still in development here but will be fully revealed in *Phaedo*).[14] Socrates takes pains to emphasize to Meno that he is not teaching but only asking questions, echoing his description of himself as a "midwife" of knowledge in *Theaetetus* (150a). This is a crucial point to Plato, for it reinforces Socrates's claim just prior to this scene that the immortal soul "has acquired knowledge of all and everything; so that it is no wonder that she should be able to recollect all that she knew before" (*Meno* 81c).

It follows from this that true knowledge cannot be taught because it is not something one person can give to another. All a teacher can offer is the techne of recollection. True knowledge resides in the soul, for Plato, and Socratic dialectic combined with geometric demonstration offered the best means by which to help humans remember what already existed metaphysically. At the same time, dialectic also exposed false belief for what it was, which can—depending on one's dependence on that belief—feel like a "torpedo's shock." Yet for those souls desirous of the truth, Plato believed, the initial numbing effect of dialectic quickly gives way to further inquiry (*Meno* 84c–85b):

SOCRATES: Now you should note how, as a result of this perplexity, he will go on and discover something by joint inquiry with me, while I merely ask questions and do not teach him; and be on the watch to see if at any point you find me teaching him or expounding to him, instead of questioning him on his opinions. Tell me, boy: here we have a square of four feet, have we not? You understand?

BOY: Yes.

SOCRATES: And here we add another square equal to it?

BOY: Yes.

SOCRATES: And here a third, equal to either of them?

BOY: Yes.

SOCRATES: Now shall we fill up this vacant space in the corner?

BOY: By all means.

SOCRATES: So here we must have four equal spaces?

BOY: Yes.

SOCRATES: Well now, how many times larger is this whole space than this other?

BOY: Four times.

SOCRATES: But it was to have been only twice, you remember?

BOY: To be sure.

SOCRATES: And does this line, drawn from corner to corner, cut in two each of these spaces?

BOY: Yes.

SOCRATES: And have we here four equal lines containing this space?

BOY: We have.

SOCRATES: Now consider how large this space is.

BOY: I do not understand.

SOCRATES: Has not each of the inside lines cut off half of each of these four spaces?

BOY: Yes.

SOCRATES: And how many spaces of that size are there in this part?

BOY: Four.

SOCRATES: And how many in this?

BOY: Two.

SOCRATES: And four is how many times two?

BOY: Twice.

SOCRATES: And how many feet is this space?

BOY: Eight feet.

SOCRATES: From what line do we get this figure?

BOY: From this.

SOCRATES: From the line drawn corner-wise across the four-foot figure?

BOY: Yes.

SOCRATES: The professors call it the diagonal: so if the diagonal is its name, then according to you, Meno's boy, the double space is the square of the diagonal.

BOY: Yes, certainly it is, Socrates.

From this last portion of Socrates's inquiry with the Boy one quickly *senses* the difficulty of following a geometric proof sans diagram. The diagram makes the geometric proof concrete and elegantly demonstrative, while foregoing the cumbersome method of exhaustion.

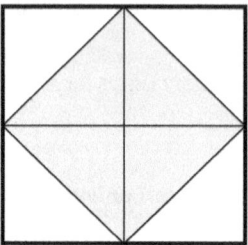

Fig. 2 | Inscribed squares

In figure 2, for instance, one can begin to grasp the meaning and significance of Socrates's words: instead of thinking arithmetically in terms of the length of the sides of the square (whether four feet or three), try to think geometrically in terms of the spatial relations between the different forms. Once one makes that shift in thinking, one stops attempting to *calculate* the length of the side and instead begins to think in terms of *halving* the square of sixteen feet, and, if one simply halves each of the squares of four feet with their diagonals, an elegant solution emerges. The shaded square above *must be* a square of eight feet not by calculation but by deduction from the accepted premise that each diagonal halves the smaller squares.

The scene between Socrates and the Boy reveals many of the central tenets of Plato's philosophy of mathematics. Note that Plato purposely chose a problem for which no arithmetic solution existed. Why? In part he did so to demonstrate the superiority of geometry and the limits of *calculation*—a second order form of mathematics for Plato.[15] Only through geometry can one "discover" the eight-foot square, and the means of discovery is patently deductive: begin with an accepted premise (diagonals halve squares) and deduce from there the *necessary* conclusion. The beauty of geometric demonstration for Plato and others of like mind comes not from the persuasiveness of the proof: the proof is not *persuasive* for Plato; it is *necessary* and *irrefutable*, proving that there must be a realm of truth beyond the everyday domains of persuasion—a thoroughly unrhetorical realm purged of controversy, debate, or argument. Plato called this realm the realm of ideas.

Equally important for Plato, the geometric proof of the eight-foot square demonstrates his principle of epistemic inquiry as recollection: simply through questioning, claims Socrates, the Boy is able to discover the eight-foot square. Now this Boy has had no previous training or education. How then is this possible, Socrates reasons, unless knowledge itself somehow already exists within him? And if within him, then where is it? Clearly not in his mind, for Socrates exposes the Boy's false belief that the side of an eight-foot square is four feet. For Plato the answer must be the soul, as he makes plain later in the *Meno* (85e–86a):

> SOCRATES: And if he did not acquire them in this present life, is it not obvious at once that he had them and learnt them during some other time?
> MENO: Apparently.
> SOCRATES: And this must have been the time when he was not a human being?
> MENO: Yes.
> SOCRATES: So if in both of these periods—when he was and was not a human being—he has had true opinions in him which have only to be awakened by questioning to become knowledge, his soul must have had this cognizance throughout all time? For clearly he has always either been or not been a human being.
> MENO: Evidently.
> SOCRATES: And if the truth of all things that are is always in our soul, then the soul must be immortal.

In this way geometric proof becomes integral to Plato's metaphysics—a means of illustrating the immortality of the soul, the discovery of truth as a process of recollection, and the need for dialectic as the midwife of knowledge.

Plato, of course, did not pluck his knowledge of mathematics out of thin air. Rather, there is strong evidence that he spent the years prior to writing *Meno* in Syracuse studying with the Pythagorean Archytas, who was renowned as a brilliant mathematician, philosopher, and statesman; there Plato took the time to seriously study mathematics, especially geometry, and it profoundly transformed his philosophy. In the dialogues prior to his studies in Syracuse, Plato's Socrates innocently performs the Socratic method of question and answer in the pursuit of *justified true belief*. However, a thinker of Plato's quality surely knew that this method could not produce *certain knowledge*, for it did not derive from *certain premises*; it was at best approximate, much like the method of exhaustion in mathematics, which we know from the scene in *Meno* could estimate but never discover the true eight-foot square. According to Gregory Vlastos, this is why we see a slow departure from the Socratic method of questioning (which he labels "elenchus") toward the more didactic, instructional form of dialectic in Plato's later dialogues.[16] The scene between Socrates and the Boy can thus be read as a kind of pivot point in Plato's thinking away from his teacher's methods and toward his own innovative form of dialectic, which was heavily influenced by Pythagorean theories of geometry.

One might find it surprising that the extent of the influence of Pythagoreanism on Plato's thinking is still coming to light in present-day research. For some time Plato scholars rejected the description of his philosophy as essentially Pythagorean (as Aristotle described it in *Metaphysics* 1.6.1, 987a).[17] In part this is because Plato hardly makes mention of Pythagoras or the Pythagoreans throughout his dialogues, nor does he at any point claim allegiance to Pythagoreanism, making Aristotle's characterization appear reductionist. Recent research, however, seems to be building toward a more nuanced view, in which the Pythagorean influences on Plato's work are becoming increasingly clear, even as that research shows how Plato's thought extended beyond and sought to improve Pythagorean philosophies.[18] Recent scholarship by J. B. Kennedy, for instance, reveals a twelve-part structure within Plato's dialogues that corresponds with Pythagorean cosmogony and music theory. As Kennedy notes, "It is in retrospect natural that Plato would have given his works mathematical form. The dialogues reflect the revolution in mathematics that affected several of the arts and sciences during the fifth century, and mathematics is thought to have been important in the early Academy."[19] One can find Pythagorean influence on Plato, then, not only in the reported trips he took to Syracuse to study with Archytas and in the increasingly prominent role of geometry in his metaphysics but also hidden in the structure of his dialogues.

Why take such pains to organize one's dialogues according to a twelve-part structure? As is often the case with Plato, the answers are many: first, we must understand that for Plato and most other Pythagoreans numbers were much more than numbers—they were pregnant with ideational meaning. The number one, for example, was less a number and more the name of the divine within Pythagorean circles.[20] And this suggests one of the central principles of Pythagoreanism, namely, that all things are number, that mathematical principles "must be the principles of all existing things" (*Metaphysics* A.5 985b23) and that mathematics has the power to disclose the divine order of the cosmos.[21] The Pythagoreans bolstered this principle with their most renowned discovery: the relationship between number and music and, especially, the alignment of whole number ratios with musical harmony. It is from these studies that the twelve-part structure comes: "According to Greek theory," Kennedy observes, "the third (1:4), fourth (1:3), sixth (1:2), eighth (2:3) and ninth (3:4) notes on the twelve-note scale will best harmonize with the twelfth." Amazingly, those sections of Plato's dialogues corresponding with harmonic notes "are dominated by positively valued concepts, while passages near dissonant notes ... are dominated by negative ones."[22] The presence of the twelve-part structure in Plato's dialogues, then, was in part organizational and in part performative—a means of articulating his metaphysics both in form and content, even as he subtly communicated his knowledge of Pythagorean mathematical theory to others within that exclusive circle.

Plato held Pythagorean mathematical cosmogony in high regard, but he did not simply inherit it; he also sought to improve it. For Plato Pythagorean metaphysics were both fragmented and incomplete: fragmented because in Plato's time Pythagoreans themselves were fragmented: some, like Philolaus, believing the objects of mathematics existed eternally; and others, like Empedocles and Epicharmus, arguing that since mathematics grows its objects too must grow.[23] They were incomplete because, for Plato, Pythagoreans had no adequate ontological theory with which to explain the power of their art. They believed that all things were number—that number was the ultimate and essential reality. In the *Phaedo*, however—where one finds a more sophisticated explication of Plato's philosophy of mathematics—we see him challenging Pythagorean orthodoxy on just this point. The dialogue takes place in Socrates's prison, where he will soon take poison in lieu of being banished from Athens. The overarching argument deals with whether willingness to die is rational or irrational, and, much like in *Meno*, Socrates uses mathematics to prove the existence of the everlasting soul and thus the rational nature of his choice. At this moment in the dialogue, Socrates is engaging with Cebes (*Phaedo* 103c–104b):

> "Well, we are quite agreed," said Socrates, "upon this, that an opposite can never be its own opposite."
> "Entirely agreed," said Cebes.
> "Now," said he, "see if you agree with me in what follows: Is there something that you call heat and something you call cold?"
> "Yes."

"Are they the same as snow and fire?"

"No, not at all."

"But heat is a different thing from fire and cold differs from snow?"

"Yes."

"Yet I fancy you believe that snow, if (to employ the form of phrase we used before) it admits heat, will no longer be what it was, namely snow, and also warm, but will either withdraw when heat approaches it or will cease to exist."

"Certainly."

"And similarly fire, when cold approaches it, will either withdraw or perish. It will never succeed in admitting cold and being still fire, as it was before, and also cold."

"That is true," said he.

"The fact is," said he, "in some such cases, that not only the abstract idea itself has a right to the same name through all time, but also something else, which is not the idea, but which always, whenever it exists, has the form of the idea. But perhaps I can make my meaning clearer by some examples. In numbers, the odd must always have the name of odd, must it not?"

"Certainly."

"But is this the only thing so called (for this is what I mean to ask), or is there something else, which is not identical with the odd but nevertheless has a right to the name of odd in addition to its own name, because it is of such a nature that it is never separated from the odd? I mean, for instance, the number three, and there are many other examples. Take the case of three; do you not think it may always be called by its own name and also be called odd, which is not the same as three? Yet the number three and the number five and half of numbers in general are so constituted, that each of them is odd though not identified with the idea of odd. And in the same way two and four and all the other series of numbers are even, each of them, though not identical with evenness. Do you agree, or not?"[24]

Cebes of course agrees, but, more important, we see the emergence in this scene of what Aristotle would later describe as "Form-Numbers" in *Metaphysics* (13.6 1080a28–35). In lieu of thinking of number as the ultimate reality like the Pythagoreans, Plato offered his theory of Forms, which named the originary ideas that unified and animated all things, including number. Thus Plato notes that each odd number shares something with all other odd numbers, and yet that which they all share is not identical with any particular iteration. There are the odd numbers, in other words, and then there is the *idea* of odd—oddness as such. There are the even numbers, and then there is the *idea* of even—evenness as such. These are the ideas behind and within numbers and they are not to be confused with any particular number nor with any particular gathering of perceptible objects. In their absorption in the arts of mathematics, Plato argued, the Pythagoreans mistook the perceptibles, whether numbers or diagrams, with the Forms themselves, failing to realize that mathematics alone cannot discover the ultimate foundations of knowledge that renders all things commensurable. For that task one needs dialectic, something Plato has Socrates explain to Glaucon fully in book 6 of the *Republic* (509d–511c):

> "You know that those who study geometry and arithmetic and similar subjects postulate odd and even, geometrical figures and the three kinds of angles, and other relationships of this sort according to each system of inquiry. So, taking these things as known, they make them their hypotheses and don't think it worth their while to offer any justification for them to themselves or others, on the grounds that they are clear to everyone. And starting from these, they go on through the remaining steps and end up in agreement at the point they set out to reach in their investigation."
>
> "Yes, of course," he [Glaucon] said, "I know all that!"
>
> "So you'll also know that they make use of the visible forms as well and make their arguments about them, although considering not the actual things, but those they resemble, making their arguments on the basis of the square itself and the diagonal itself, but not the line they are drawing, and similarly with everything else. These very things they are forming and drawing, of which shadows and reflections in water are images, they now in turn use as their images and aiming to see those very things which they could not otherwise see except in thought."
>
> "You're right," he said.
>
> "This then is the class that is intelligible ... where a soul is forced to use hypotheses in its search for it, without working toward a first principle because it is unable to escape from its hypotheses to a higher level, but by using as images the very same things of which images were made at a lower level and, in comparison with those images, were thought to be clear and valued as such."

"I understand that you mean geometry and those arts related to it."

"So understand, too, what I mean by the other section of the intelligible, which reason itself grasps by the power of dialectic, using hypotheses which are not first principles, but genuine hypotheses, like steps and starting points, in order to go as far as what is unhypothetical and the first principle of everything. And, grasping this principle, it returns once again, keeping hold of what follows from it, and comes down to a conclusion in this way, using no sense perception in any way at all, but Forms themselves, going through Forms to Forms and ending up at the Forms."

I cite Plato at length here for the last time, kind reader, for it is here at last, in the *Republic*, that Plato arrives at his most complete explication of his philosophy of mathematics, a philosophy we have seen slowly evolve since the scene between Socrates and the Boy. In contrast to *Meno*, where geometry is an idealized method of recollection of knowledge, here Plato is much more circumspect: mathematics is certainly still important, retaining its prominent place in philosophical education, but alone it fails to bring one to the realm of ideas.[25] Mathematicians taken with the elegance of their geometric demonstrations remain at the level of hypotheses, not thinking "it worth their while to offer any justification for them [basic hypotheses] to themselves or others." Thus, for Plato, they fail to attend fully or adequately to the first principles of their art, that which unifies and renders all mathematics commensurable with itself and all things commensurable with mathematics, properly understood. For that, one needs dialectic, which transcends the limits of hypothesizing in thinking with the "Forms themselves, going through Forms to Forms and ending up at the Forms."

Much is at stake here for Plato: the nature of knowledge as unchanging and thus the ultimate order underlying all apparent disorder (whether random change, luck, or chance); the hierarchical relationship between dialectic and mathematics; and the place of mathematical objects in Platonic metaphysics. It is clear from the previous passage that for Plato mathematical objects are not ideas themselves but animates of the ideas, media through which the Forms reveal themselves, enabling the (re)collection of *aletheia*. The Forms shine more brightly through number and geometry than through most other perceptibles (thus the role of mathematical training in the philosopher's life), yet both numbers and geometric diagrams remain perceptibles. Through dialectic, however, one can begin to think beyond numbers and diagrams to the Form-Numbers themselves; that is, one can learn to think not of the number three but of threeness as such. Only then will one begin to understand why Plato spent so much time discussing the number three (most prominently in *Phaedo* and the *Republic* but elsewhere too): it has contained within it oneness, evenness, and oddness—the three Form-Numbers that are the essence of number itself, that which is necessary to construct all numbers.[26] When one employs dialectic, then, one departs the realm of appearances, becoming, and change and enters the realm of ideas, being, and permanence, the only place where the true essence of unity and division, justice and injustice, can be thought unfettered by "sense perception."

At the start of our journey, then, Plato apotheosized geometric demonstration as a means of establishing the immortality of the soul and his theory of knowledge's discovery as a form of recollection. Yet he was still committed to the Socratic method of questioning (*elenchus*), which comes with a certain inherent respect for the other. Then in the *Phaedo* we see Plato pushing beyond his Pythagorean forbearers with the notion of Form-Numbers, which reside in an ambiguous place between the realm of appearances and ideas, respectively, and with this shift we see a corresponding departure from Socratic inquiry.[27] Finally, by the time he wrote the middle books of the *Republic*, Plato's metaphysics had matured and hardened such that the study of the mathematical arts (calculation, geometry, and harmonics) were placed in a prominent but secondary role to dialectic, a hierarchy that seems to align with the dogmatic thrust of the *Republic*, whose dialogic style serves as a thinly veiled veneer for its authoritarian content.

Equally important for understanding Plato's philosophy of math, his weaving of mathematics and dialectic in the *Republic* was simultaneously a weaving of mathematics and ontology. For in thinking dialectically about mathematical objects, one "naturally," according to Plato, arrived at the higher Forms—Form-Numbers (for arithmetic) and Platonic Forms (for geometry)—that make up the essence of the *ars mathematica*. And for Plato the only way to ultimately justify the eternal and unchanging existence of mathematical objects was to establish a philosophical theory equal to the task. As we will see in the following section, the consequences and influence of Plato's ideas regarding mathematics are difficult to overstate, especially for those interested in understanding math as a symbolic, discursive, argumentative, rhetorical practice.

Plato and His Discontents

Plato's treatments of mathematics in *Meno*, *Phaedo*, and the *Republic* serve as fascinating dialogic moments not merely for the elements of his philosophy we can piece together from them but also for the evolution of thinking that emerges through their juxtaposition. That evolution points to one of the major consequences of Plato's philosophy of mathematics, namely, the concealment of the materiality of mathematical discourse. The scene between Socrates and the Boy is remarkable in part because of the conspicuous presence of math's thick discursive materiality—its habits of diagramming and inscription that allow Socrates to *guide* (Socrates's claims notwithstanding) the Boy toward the geometric demonstration of the eight-foot square. The scene between Socrates and Glaucon in the *Republic* is equally remarkable for the conspicuous absence of such materiality. In *Meno* math's materiality appears crucial as a means of proof of the immortality of the soul and the mnemonic theory of knowledge. Recall, for example, all the points where Socrates inscribes, draws, and describes for the Boy. By the time Plato writes the middle books of the *Republic*, however, math's materiality appears incidental, perhaps even an obstacle to thinking the Form-Numbers as such, and the relative absence of representations of mathematics as a material practice are correspondingly few. In *Meno* geometry stands on its own feet—Plato weaves it in with his metaphysical theories, but it remains independent and autonomous, and Plato does not seek to subordinate it. In the *Republic*, by contrast, mathematical learning is placed fully in service to philosophical education and is ultimately subordinate to the study of dialectic.

The slow subordination and marginalization of math's discursive materiality in Plato's dialogues is not coincidental but elemental to his philosophy. For Plato discourses, languages, and symbols are always images, representations, or imitations of the true Forms that animate them. Plato says as much explicitly in *Cratylus* (432b–c), where he has Socrates explain that an image that reproduced all the qualities of "that which it imitates" would cease to be an image and instead become the thing itself.[28] The thing would be duplicated rather than imitated. This logic of *mimesis* as derivative—or what Phillip Sidney Horky calls Plato's "principle of deficiency"—permeates all dimensions of his philosophy of language.[29] Words and symbols, especially written discourse, are mere images of the Ideal Forms that animate them, thus always derivative of those Forms and burdened by an inherent, constitutive lack. Of course, that means words and symbols can never *be* Truth. At their best, however, they can operate as vehicles of Truth's recollection and dissemination. To those readers familiar with Plato's *Gorgias* and *Phaedrus*, these ideas about words and symbols will sound eerily familiar, for they are the very same ideas that Plato uses to initially condemn rhetoric (in *Gorgias*) and then eventually rehabilitate it as a *potential* (though dangerous) ally of truth (in *Phaedrus*).

One can already detect the influence of Plato's broader philosophy of language on his treatment of geometry in *Meno*, where one finds Socrates using an *unlettered* diagram to guide the Boy. On its own this fact appears unremarkable, but by the time Plato wrote *Meno* the use of lettered diagrams was conventional among Greek geometers, suggesting the absence of letters was deliberate.[30] Why would Plato purposely omit letters in his geometric demonstrations (not just in the scene between Socrates and the Boy but elsewhere as well)? And why would discussion of specific geometric diagrams diminish in the dialogues as Plato progressed? Reviel Netz, in his insightful treatment of the practices of ancient Greek geometry, offers two reasons. First, Plato's dialogues filtered out these diagrams in part because they were meant to reflect conversation "so that diagrams used by the speakers must be reconstructed from their speeches." Second, the filtering out of physical models, whether diagrams in the case of geometry or planetaria in the case of astronomy (see *Timaeus* 40d), aligned with Plato's theory of Forms and his belief that perceptible objects are useful only to a point (see the earlier discussion of the *Republic*).[31] To these reasons I would add the following:

1. Due to this doubled filtering, the actual practices of doing math are largely absent from Plato's dialogues. We know this was not from lack of ability, suggesting that what Plato does include is perhaps meant as enticement: for those who want to learn more, come to the academy;
2. Plato's displacement of the material practices of mathematical inscription with dialogue was yet another performative act—a means to encourage if not force his readers to prize dialectic over all other modes of inquiry, including mathematics;
3. The filtering strategy allows Plato to use discursive form to reinforce philosophic content—just as the Forms animate all appearances, so the mathematical Forms discreetly animate Plato's dialogues; and
4. Plato's philosophy of language ultimately reduces physical models and the practices of inscription in mathematics to derivative representations—not the real objects of mathematical thought but inevitably flawed imitations.

As a result, Plato never engaged and likely underestimated the constitutive role of inscription, not just in the communication of math but also in the processes of mathematical concept formation.

Ironically, then, even as Plato believed one must ultimately seek to transcend language and symbol, he simultaneously reduced diagrams and models to signs of the higher Forms they supposedly represented. And this reduction of diagrams and models to representations is one of the more powerful engines behind the widespread notion of mathematics as abstract thought. For it is precisely out of Plato's imposition of a theory of representation on mathematical practice that at once creates an alienating distance between mathematical objects and those who think them and requires the notion of abstraction to explain the relationship between mathematicians and the alien mathematical objects they attempt to think. And, if all mathematical inscriptions are but representations of real mathematical objects that exist independent of those who think them, then the whole enterprise of doing mathematics shifts: instead of focusing on the arts of innovation (inscription, diagramming, conjecture, argumentation), one focuses on the arts of justification (absolute truth, deductive logic, stable definitions, Ideal Forms); instead of thinking in terms of plurality and creativity, one thinks in terms of similitude and correctness. It is the correctness of the representation to its true Form that matters, not the innovative style of the expression.

The arts of innovation and justification are not mutually exclusive for Plato, nor are they for any other sophisticated mathematical realist, but Plato's tendency is toward justification and unity and commensurability. Plato's philosophy of math, then, is much more dogmatic than that of his Pythagorean predecessors. There are not many maths but one math; there are not many circles but one true circle; there are not many harmonies but one *harmonia*. As he states explicitly in *Epinomis* (991d–992a), "To the person who learns in the right way it will be revealed that every diagram and complex system of numbers, and every structure of harmony and the uniform pattern of the revolution of the stars are a single thing in concord with all these phenomena. And it will be revealed to anyone who learns correctly, as we say, fixing his eye on unity. To one who studies these subjects in this way, there will be revealed a single natural bond that links them all."[32] A kind of intellectual dogmatism emerges from this philosophical perspective (there is one "right way" to learn, one true definition, one unity), and geometry was the most powerful instrument Plato had for its broad imposition. Geometry was that which most clearly appeared to escape ambiguity and controversy with *apodeictic* propositions; and, consistent with Plato's monist metaphysics, one can see him imposing a deductive geometric logos on all realms of human activity, from music and the arts to medicine and gymnastic to rhetoric and politics.

Plato's desire to render all things commensurable emerges clearly in *Protagoras*, an early dialogue in which he engages the problematic around the forces of *tuchē* (luck, chance) and *technē* (art, science). There Plato ultimately argues for a *technē* of measurement, which for him promised humans both a modicum of control and the means to escape the unpredictable whims of *tuchē*. As Martha Nussbaum notes in her incisive analysis, for Plato the only force that will save humanity from the warps and woofs of *tuchē* is "something that will assimilate deliberation [and the values therein] to weighing and measuring; this, in turn, requires a unit of measure, some external end about which we can all agree, and which can render all alternatives commensurable. Pleasure enters the argument as an attractive candidate for this role." But, as Nussbaum later clarifies, whether pleasure or some other ultimate good, the problem for Plato in *Protagoras* and elsewhere was "to find *a* standard or measure that will render values commensurable, therefore subject to precise scientific control. . . . What we need to get a science of measurement going is, then, an end that is single (differing only quantitatively): specifiable in advance of the *technē* (external); and present in everything valuable in such a way that it may plausibly be held to be the source of its value."[33] If we can just determine the true Form of all values, thought Plato, we can use that *knowledge* to inform specific conflicts between values, enabling us to make *rational* decisions within those contexts. With knowledge of that which informs all values, Plato thought a science of judgment would become possible.[34] No longer will we be mired in unresolvable arguments over seemingly incommensurable values, nor will we need the democratic institutions of persuasion required to manage such arguments.

Geometric demonstration was a powerful ally in Plato's drive toward commensurability. Combined with dialectic, it not only offered "evidence" for his metaphysical theories but also provided a *correct* model for thinking and engaging with problems—not just problems related to mathematics, mind you, but *all problems*.[35] And here we begin to see the final and perhaps most significant consequence of Plato's philosophy of math for this book, namely, its dogmatic imposition on all dimensions of human activity and the resultant alienation and displacement of other ways of knowing (a phenomenon still very much alive in contemporary culture). This is why, in *Gorgias*, Plato's famous polemic against the Sophists and their *rhêtorikê*—a topic seemingly far removed from the interests of math—one finds a geometric logos at work.[36] In *Gorgias* Plato does not deploy geometry as he does in *Meno*, nor does he even discuss it at length as he does in *Phaedo* or the *Republic*. No, his use of geometry is much more subtle in *Gorgias*, which is perhaps why so few scholars have mentioned it.[37] Yet, as we have noted throughout this chapter, Plato's middle and late dialogues almost always contain a mathematical substructure. In *Gorgias* that substructure manifests in a dramatic battle between the two great inventions of ancient Greece: geometry and democracy.

Given Plato's metaphysical commitments, one can imagine his impatience with the demos, mired as it was in intractable arguments that for Plato were a result of ignorance (sometimes willful ignorance). The Sophists were emblematic of all that was wrong with democracy, and so Plato dramatized an encounter between Socrates and several of the more prominent Sophists of Athens. The dialogue begins with Gorgias, one of the most famous Sophists of his generation. As usual, Socrates starts the dialogue with a deceptively simple request: to ask Gorgias "what is the function of his art and what it is that he professes and teaches" (*Gorgias* 447c). Gorgias responds that he is a rhetorician (449a) and that rhetoric is the art of speech. Socrates, being Socrates, probes on: surely Gorgias does not mean all speech, for, "one might retort, if one cared to strain at mere words: So, Gorgias, you call numeration rhetoric! But I do not believe it is either numeration or geometry that you call rhetoric" (450e). In the very earliest moments of the dialogue, then, we see our first "chance" meeting between rhetoric and mathematics. Surely, Socrates implies, it would be absurd to consider mathematical speech rhetorical, to which the straw Gorgias replies, "Your belief is correct, Socrates, and your retort just" (451a). Hints of an emerging chasm between rhetoric and mathematics are thus already forming in the first pages of Plato's dialogue, the former dealing with "mere words" and the latter with something else entirely. Even so, Plato positions math as the model on which the inquiry into rhetoric should proceed: if someone asks, "what is the art of numeration? I should tell him, as you did me a moment ago, that it is one of those which have their effect through speech. And suppose he went on to ask: With what is its speech concerned? I should say: With the odd and even numbers, whatever may chance to be the amount of each. . . . Come then and do your part Gorgias. . . . What subject is it, of all in the world, that is dealt with by this speech employed by rhetoric?" (451a–d).

So early in the dialogue and Plato has already set his terms: the inquiry is supposedly into the nature of *rhêtorikê*, and yet math has already been established as nonrhetorical and, ironically enough, the model to which the inquiry into rhetoric should aspire. Math has objects of study (numbers, geometric figures) and first principles (of numeration, of calculation) to teach; it offers knowledge of those objects to those who study it and can thus be rightly considered an art (*technê*). Using this model definition of *technê*, Plato proceeds to examine rhetoric and, not surprisingly, finds it wanting. Through Socrates's questions it is revealed that rhetoric as practiced by the Sophists has no clear object of study, thus no first principles to impart and no knowledge to offer. After being pressed by Socrates, Plato's Gorgias clarifies that rhetoric is speech dealing "with what is just and unjust" (454b) and yet Socrates shows that Gorgias has no knowledge of true justice. How can one *correctly* persuade an audience, Plato implies, toward the just and away from the unjust without knowledge of the nature of justice? Many Sophists claim to teach virtue, but for Plato that is absurd because, as Socrates shows, they have no knowledge of true virtue nor even believe such a thing as true virtue exists. In Plato's hands Sophistry is reduced to *eristic* rhetoric, a false art that impersonates the true art of legislation through the machinations of *"flattery"* (463a). Enter straw Callicles, who embraces rhetoric's *eristic* virtue—its capacity to bring power to those who wield it well and thus promote the natural order in which the strong dominate the weak (482c–486d). All else—laws, values, beliefs—are mere conventions for Plato's Callicles, obstacles to the rare individual who "bursts his bonds, and breaks free . . . tramples underfoot our codes and juggleries, our charms and 'laws,' which are all against nature" (484a).

By the end of *Gorgias*, then, Plato has reduced rhetoric to a phantom, a false art taught by false prophets who at best know not what they do and at worst have a hedonistic desire for pleasure and power. To Callicles's apotheosis of *eristic* rhetoric and ridicule of philosophy as esoteric theory babble, Socrates offers a revealing retort (*Gorgias* 507e–508a): "And wise men tell us, Callicles, that heaven and earth and gods and men are held together by communion and friendship, orderliness, temperance, and justice; and that is the reason, my friend, why they call the whole of this world by the name of order [cosmos], not of disorder or dissoluteness. Now you, as it seems to me, do not give proper attention to this, for all your cleverness, but have failed to observe the great power of geometrical equality amongst both gods and men: you hold that self-advantage is what one ought to practice, because you neglect geometry." And here we finally see the dialogue explicitly return to mathematics as the model against which other intellectual forms can and should be measured. It is from the study of geometry that one learns of the inherent order of the cosmos and thus knows better than to accept Callicles's argument, which ultimately depends on the random luck (*tuchē*) of birth. Here we also see more clearly than at most other points in his dialogues how Plato imposed a geometric logos on other domains—in this case on both rhetoric and the demos.

Democracy lives only in its expressive rhetorical forms. But for each of these forms (oratory, assembly, argument, theater, sculpture, architecture) and each of their positive capacities for allowing a collective to represent itself to itself as a collective, Plato substitutes the following mathematical principles: (1) for the democratic model of translation in which citizens have the capacity to *act*—to interpret events and invent arguments to advance their interests—he substitutes a geometric model of diffusion, where truths are spread to the people and where citizens might even be forced, if necessary, "to adopt a course of action which would result in their becoming better people" (517c). (2) For the democratic practice of collective demonstration in which the polis can express itself to itself as a *citizen-state*, he substitutes the logos of geometric equality, which, as Bruno Latour notes, "requires a strict conformity to the model since what is in question is the conservation of proportions through many different relations."[38] And (3) for the dynamic urgency of politics—and the invention of rhetorics equal to it—he substitutes the statics of an axiomatic mathematical model of language in which statements demonstrated as true become the benchmarks against which all others are judged.

Implications

From *Gorgias*, then, we can finally see how Plato rendered rhetoric and mathematics oppositional intellectual enterprises, a view that persists widely to this day. We can also see how Plato used a geometric logos to diminish rhetoric and undermine democratic politics. But in Plato's other dialogues, one sees an interesting progression, from the particularities of mathematical practice to the generalities of mathematical forms. And herein lies at least one source of Plato's mythos of math: namely, his increasing emphasis on forms and commensurability over empirical practice. For it is in the turning away from mathematical practice—from that embodied act of scribbling, thinking, imagining, and diagramming—that one can begin to forget math's rich discursive materiality and begin to shroud that same practice in a mystifying transcendental metaphysics. Unfortunately, the consequences of such mystification are significant, for not only is respect and appreciation for the arts of the demos diminished, mathematics too is transformed—away from math as an art of creative thinking and innovation and toward math as an abstract form of absolute truth, a derivative of which is the axiomatic practice of instrumental problem-solving.

Equally unfortunate, much contemporary work testifies to the continued influence of Plato's ideas in the twenty-first century: Mark Balaguer's study of different philosophies of math, for instance, found that mathematicians "sometimes use the term 'realism' interchangeably with 'platonism' . . . because it is widely thought that platonism is the only really tenable version of realism."[39] In one of the few surveys of practicing mathematicians' views of math, Leone Burton concluded that "mathematicians do mostly subscribe to an absolute view of mathematics."[40] And numerous studies of mathematics education link realism with an axiomatic approach in the classroom, which tends to alienate students while communicating that mathematics is an abstract language of truth primarily about rule-driven problem-solving.[41] Beyond the confines of mathematical communities, however, the consequences are even more severe, with mathematical models increasingly treated as infallible technologies rather than human-designed discursive constructions full of assumptions and domains of validity. Mathematics as a result often functions in the public sphere as a cudgel stifling discussion instead of as a catalyst for further thought and insight.[42]

While the power and influence of mathematical realism remains formidable, in the past few decades, an increasingly broad chorus of voices have started to challenge the mythos of math. As one can see in chapter 2, each voice captures a piece of what is a broader rhetorical approach to mathematics, one that embraces mathematical discourse and practice neither as absolute truth nor constructivist fiction but as an increasingly powerful *translative rhetorical force* that we must come to understand if we, as a species, are to successfully navigate our own growing capacities for domination, control, and destruction.

2

Imbrications | Mathematics as a Translative Rhetorical Force

Rhet-o-ric: language designed to have a persuasive or impressive effect on its audience, but often regarded as lacking in sincerity or meaningful content.
Math-e-mat-ics: the abstract science of number, quantity, and space.
—Oxford English Dictionary

The *Oxford English Dictionary*'s definitions of rhetoric and mathematics testify to an old and persistent polarization. Rhetoric—associated with the use of language to persuade, influence, and manipulate—plays the intellectual counterpart to mathematics: the search for abstract, pure, infallibly true relations between number, geometry, and space. This polarization, originally codified in Plato's dialogues, pervades contemporary culture. The influential twentieth-century mathematician and philosopher Bertrand Russell defined math as that which, "rightly viewed, possesses not only truth, but supreme beauty—a beauty cold and austere, like that of sculpture, without appeal to any part of our weaker nature." He later expanded his praise: "Mathematics takes us still further from what is human, into the region of absolute necessity, to which not only the actual world, but every possible world, must conform."[1] In contrast, Russell characterized rhetoric as "the appearance of great wisdom . . . [but] the reality of witchcraft."[2] More recently, Philip Davis and Reuben Hersh gave voice to the aporia between rhetoric and mathematics: "Mathematization means formalization, casting the field of study [any field of study] into the axiomatic mode and thereby . . . purging it of the taint of rhetoric."[3] One need not look far to find assertions about rhetoric and math that fall along these lines, scoring and then scoring deeper the chasm that renders them ever more distant strangers. So opposed do these strangers appear today that most—both inside the academy and out—see them as antithetical.

Yet, in the second half of the twentieth century, the story of rhetoric and mathematics' long estrangement began to shift. Slowly but surely scholars began to question the role of mathematics in matters political, wondering how math is used to persuade, questioning the infallibility of even rigorously proven mathematical statements, and suggesting that perhaps math does not simply apprehend the real—perhaps it expands and transforms the real. In tracing these efforts, not only do we see why and how contemporary thinkers have challenged Plato's views, but we also begin to see an exciting alternative perspective form, one that requires us to revise our understanding of both rhetoric and math to see how the two have, despite Plato's claims, worked collaboratively all along.

What is at stake in the campaign to rebuild commonplaces between rhetoric and math is nothing less than a major transformation in thinking about rhetoric, mathematics, and culture. What previous scholarship shows time and again is that when one attends closely to the practices of doing math, one is consistently confronted not by a priori truth but by the slow, deliberate, painstaking fabrication of complex discursive networks out of which, on occasion, a principle of composition emerges, one with the translative rhetorical force necessary to materially transform reality. This perspective, I argue, not only helps us see mathematics and rhetoric anew but also helps us better understand our increasingly powerful capacities as humans to rewrite (often in ignorance) our own social-material worlds.

Circumventions

Plato's machinations in his dialogues would be laughable if they were not so tragically influential. I need not rehearse for contemporary scholars how Platonic thought was extended, adapted, transformed, and hardened with the rise of modernism, nor how it continues to haunt contemporary culture.[4] As Brian Rotman, a mathematician who studies the semiotics of math, notes, "For most mathematicians (and, one can add, most scientists) mathematics is a Platonic science, the study of timeless entities, *pure forms* that are somehow or other simply 'out there,' preexistent objects independent of human volition or of any conceivable human activity."[5] Rotman's work figures as an early and sophisticated effort to develop an alternative to Platonic realism and thus serves as a productive starting point for understanding how Platonic thought gets rearticulated within modern mathematical discourse as well as how it obscures the actual practice of mathematics.

The Semiotic Approach

The organizing principle of Platonic realism is that mathematical objects exist independently and a priori of human cognition. Mathematical symbols are, accordingly, more or less adequate representations of ideal mathematical objects, and the purpose of doing mathematics is to discover "objective irrefutably-the-case descriptions of some timeless, spaceless, subjectless realm of abstract 'objects.'" From this perspective mathematical symbols should function as transparent referents for real mathematical objects and their relationships and should have no function beyond that: "Language, for the realist, arises and operates as a name for the preexisting world. Such a view," Rotman observes, creates "a bifurcation of linguistic activity into a primary act of reference—concerning what is 'real,' given 'out there' within the prior world waiting to be labeled and denoted—and a subsidiary act of describing, commenting on, and communicating about the objects named."[6] Thus the hard Platonic division between the work of representation and invention: when mathematicians assemble an equation that proves valid, they are not inventing; they are *describing* and *discovering*.

While this perspective may have some psychological benefits—enabling those operating from within it to see their work as pure and absolute—it also has some real deficits when it comes to understanding mathematical practice. Like the realist style so familiar to scholars in other domains, mathematical realism immediately denies the materiality of mathematical discourse.[7] True mathematical symbols, formulas, and theorems are not interpretations or even representations of an ideal mathematical reality for realists; they are transparent *presentations* that, taken together, "merely *describe* prior mental constructions appearing as presemiotic events accessible only to private introspection." But how, asks Rotman, can we account for the movement from mathematical practice to mathematical knowledge? By what means do mathematicians capture their supposedly presemiotic thoughts? And what is the mathematician's relationship with the a priori realm of mathematical objects?[8] Plato answers with reference to the soul and a metaphysics of reincarnation; Gottlob Frege, one of the founders of formal logic and a preeminent mathematical realist of the twentieth century, does not do much better: "The apprehension of a thought presupposes someone who apprehends it, who thinks it. He is the bearer of the thinking but not of the thought. Although the thought does not belong to the thinker's consciousness yet something in his consciousness must be aimed at that thought. But this should not be confused with the thought itself." For Frege there are objective, timeless thoughts and there are thinkers who can "apprehend" them, but the "something in his consciousness" that allows the finite human subject to access the infinite realm of absolute truth remains enigmatic.[9]

For Frege and Plato and other mathematical realists, Rotman argues, the means of production of mathematical knowledge are hidden because the constitutive work of the mathematical sign is systematically denied. Taken together Rotman's three books—*Signifying Nothing*, *Ad Infinitum*, and *Mathematics as Sign*—figure a semiotic intervention into the space between rhetoric and mathematics that Plato originally opened and that largely remains unexplored today.[10] Building on Charles Sanders Peirce's semiotic work, Rotman's semiotics suggest that the first step to understanding math as a discursive practice is to ask how mathematical discourses subjectify those who practice them. When you do mathematics, in other words, who are you? Two primary roles, Rotman argues, immediately emerge in mathematical discourse: "the one who imagines (what Peirce simply calls the 'self' who conducts a reflective observation), which we shall call the *Subject* and the one who is imagined (the skeleton diagram and surrogate of this self), which we shall call the *Agent*."[11]

Initially, the relationship between the Subject and the Agent appears straightforward—not unlike someone who uses a map, projecting an imaginary version of themselves into the mapped space, but, unlike a map, mathematical signs signify a "purely imaginary territory." This imagined space—and the differences it introduces—is crucial to understanding mathematical practice: when one thinks mathematically (add all odd numbers between one and one hundred, for instance) one simultaneously projects oneself into an imagined world (in our example, a world where such "things" as numbers and number lines "exist") and then executes the necessary operations. Crucially, however, this practice of thinking and imagining and executing is, Rotman observes, constituted, buttressed, and reinforced by an equally important semiotic practice: "Such creation cannot ... be effected as pure thinking: signifieds are inseparable from signifiers: in order to create fictions, the Subject scribbles."[12]

That scribbling—that material, corporeal, semiotic element of mathematical practice neglected by realists (recall the progressive concealment of inscription in Plato's dialogues)—becomes even more important when our mathematical thinking moves from the finite (as in our previous example) to the infinite (add the series 1, ½, ¼ . . .), for at that point we must not only project ourselves as a Subject into an imaginary mathematical world but also imagine an Agent unrestrained by "finitude and logical feasibility—he can perform infinite additions, make infinitely many choices, search through an infinite array, operate within nonexistent worlds."[13] Here the support of written signs becomes essential not simply as a recording device for one's a priori mathematical thoughts but as a *means of mathematical innovation*: when we attempt to solve new or unfamiliar problems, for instance, we use signs to marshal our previous knowledge (both mathematical and nonmathematical) *and* we use signs to create nonfinite thought-experiments that the Subject can, via the Agent, test.

Although we have only scratched the surface, two important insights emerge from Rotman's work at this point: first, the structure of mathematical persuasion within the practice of doing mathematics begins to take shape, and, second, the forms mathematical discourse takes and their consequences begin to emerge. Let us deal with these in turn: if, as Rotman suggests, mathematics is a series of thought-experiments in which mathematicians imagine themselves into a mathematical world and, via inscription, make predictions that they test through their imagined Agent, how are they persuaded (mathematicians would say compelled) to believe in the integrity of their testing? Unlike the physical sciences, mathematicians have no physical world with which to verify their predictions. By what means are they compelled? By "observing" the "actions" of the imagined Agent projected into the already established mathematical world, Rotman argues, the Subject is persuaded that the predictions within the thought-experiment are probable. However, it is through a more rigorous proof procedure that mathematicians are convinced of a statement's absolute truth. That proof procedure is composed of both fictive and logical elements—the fictive figure of the imagined Agent and the logical force of the scribbling Subject. Each layer of a mathematical proof, Rotman claims, sustains a dialectical tension between these elements, even as each layer corresponds with the scribbling and manipulating of written signs: "These manipulations form the steps of the proof in its guise as a logical argument: any given step either is taken as a premise, an outright assumption about which it is agreed no persuasion is necessary, or is taken because it is a conclusion logically implied by a previous step."[14]

Yet this description of mathematical persuasion is inadequate, for animating every nontrivial mathematical proof is what Rotman calls a "leading principle," an idea that motivates the logical minutiae of a proof: "Presented with a new proof or argument, the first question the mathematician . . . is likely to raise concerns 'motivation': he will . . . seek *the idea behind the proof*. He will ask for the story that is being told, the narrative through which the thought experiment or argument is organized."[15] Why, one might ask, are these ideas absent from the mathematical proofs they animate? Because they exceed the strict boundaries of formal logical discourse sanctioned within formal proofs (what Rotman calls "the Code").[16] Rotman's argument here, and other scholars have made a similar point, is that one can know a proof, reproduce all the steps, even explain why those steps follow logically from one another, and yet fail to *understand the meaning and significance* of the proof.[17] That is because meaning and significance come from "*the idea behind the proof*," and those ideas exist only semiotically in the "meta-Code"—that is, in the discursive formations excessive to formal logic. This aspect of mathematical practice forces Rotman to add a third figure to his semiotic model: the Person. "The Person constructs a narrative, the leading principle of an argument, in the meta-Code; this argument or proof takes the form of a thought experiment in the Code; in following the proof the Subject imagines his Agent to perform certain actions and observes the results; on the basis of these results ... the Person is persuaded that the assertion being proved—which is a prediction about the Subject's sign activities—is to be believed."[18] The Person—the actual corporeal, physical presence who scribbles and thinks, thinks and scribbles, and, most important, has access to the ideas in the meta-Code—names the motivational force behind formal mathematical proofs. Now we have a more complete semiotic picture of mathematical persuasion.

It is precisely the Person—who is finite, lives outside the formal mathematical code, and has access to the ideas that motivate mathematical theorems and proofs—who Platonic realism denies. That denial brings us to our second insight: although Platonism protects a faith in mathematics as absolute, it also shapes conventional mathematical discourse in profound ways. Consider the conspicuous absence of indexical terms (*I, now, here*), for instance, in most mathematical texts; the omnipotent voice of command (the imperative mode); or the lack of contextual explanation of the concepts considered.[19] Each of these discursive characteristics echoes an investment in a Platonic worldview: the absence of pronouns reinforcing realism's basic principle (math is about real objects), the imperative mood aligning with a hierarchical theory of Forms, and the ahistorical style fitting with claims to transcendental truth. Yet if one desires to communicate the actual practice of doing (rather than memorizing and regurgitating) math and hopes to offer an understanding of the meanings of the concepts mobilized in mathematical statements, then one needs to attend to the creative, constitutive force of mathematical discourse that Platonic realism renders invisible and that only now is beginning to come into view through Rotman's analyses.

For rhetorical scholars the good news is that several old friends traffic with Rotman's reanimation of mathematical discourse: the meaning and significance of a mathematical statement comes not from its infallibility—the establishment of which is the purpose of formal logic—but from its meta-Code. Suddenly we see that contextual knowledge—rhetorical, historical, sociopolitical—of mathematical concepts is central to *understanding* them, without which one can execute a problem or a proof (as a computer might) and yet not have the slightest inkling of the meaning or purpose of that execution.[20] Equally problematic, one who encounters Platonically inspired mathematical discourse is likely to see it as an inert, abstract, rule-driven system of formal logic instead of as a fascinating, evolving, contextually situated, creative practice of thinking and writing. The *arts of mathematical innovation* lie in the nexus between thinking and scribbling and the argumentation that these "thought-scribbles" give rise to within one's mathematical community.[21] And this, finally, is why I describe Plato's view as a mythos *of* mathematics in the previous chapter: it shrouds many of the most fascinating aspects of *mathematical practice* behind a metaphysical veil that is both damaging—to math, to rhetoric, to politics—and, Rotman shows us, unnecessary.

An attentive reader might reasonably ask at this point, but how does Rotman explain mathematics' usefulness in the physical sciences? To parrot the physicist Eugene Wigner, how can one explain the "unreasonable effectiveness of mathematics" without reference to the a priori?[22] Math's effectiveness is only "mysterious," Rotman rejoins, if one already sees with Platonic eyes—that is, sees a world where finite humans have access to and can discover eternal truths and where the Person and the meta-Code critical to *mathematical practice* are invisible. Rotman's alternative proposition is that "mathematical objects are not so much 'discovered out there' as 'created in here,' where 'here' means the cultural circulation, exchange, and interpretation of signs within an historically created and socially constrained discourse." This does not amount, Rotman emphasizes, to a reductive social constructivism; instead, there is a "twofold movement between mathematical signifieds and the world."[23] Each constitutes and is shaped by the other in an evolving series of transformations: "mathematical signifieds themselves owe their origin to empirical, material features of the world," and the world, especially the techno-scientific world of late capitalism, owes its shape and influence to the forms mathematical signifieds often give to it.[24]

The semiotic program Rotman initiates is sophisticated and productive, developing a semiotic theory of mathematical practice that reveals the structure of mathematical persuasion; offering a grounded, nonmetaphysical account of math that challenges Platonic realism; and opening a whole realm of symbolic action to rhetorical analysis. Yet Rotman's approach, like any analytic, has limitations. Rotman's work has many fascinating implications, but often those implications are left undeveloped. His description of mathematical statements as predictions, for example, suggests mathematical discourse might be largely deliberative in character, but analysis of this possibility is absent from his work. Likewise, Rotman shows that mathematical meaning comes—through the Person—from the meta-Code external to formal mathematical discourse. But what are the analytic strategies available to scholars—short of talking to the Person—to tease out the different ways meanings get produced in mathematics, to study how they circulate, and to analyze how they influence mathematical practice? Perhaps a result of his intense focus on mathematical signs is a certain kind of shortsightedness when it comes to larger macro questions. Fortunately, other scholars have begun to address some of these broader issues, building on and extending Rotman's work.

The Cognitive-Metaphorical Approach

Were it not for a shared interest in mathematics, George Lakoff's and Rafael Núñez's *Where Mathematics Comes From* could not be more different from Rotman's work. In their book one finds a more quotidian style, one less given to philosophical polemics, and, while they have an interesting way of sidestepping Platonic realism, unlike Rotman they do not appear compelled to challenge it directly. Likewise, their method of reading mathematical discourse—a combination of linguistics, cognitive science, and metaphoric analysis familiar to readers of Lakoff's previous work—puts Rotman's semiotic approach in sharp relief, if only for its seemingly linear execution. Yet several affinities emerge: both show an inclination toward close analysis of mathematical discourse, both have an abiding interest in the ideas that motivate mathematics, and both conclude that mathematical *practice* cannot be accounted for from a strict Platonic perspective. Their differences, however, are conspicuous, and in attending to them we can glimpse the complexity of the relationship between rhetoric and mathematics. It is not so much that their projects are incommensurate. The relevant distinction, rather, comes from how each conceives of the production of mathematical meaning.

Where Mathematics Comes From represents an ambitious effort to begin a new area of study the authors call "mathematical idea analysis." They immediately set their project apart from Rotman's: "The intellectual content of mathematics lies in its ideas," they claim, "not in the symbols themselves."[25] Thus, for Lakoff and Núñez any account of the origins and practices of mathematics must attend to mathematical ideas, and if those ideas emerge from human cognition, as they argue, then a cognitive scientific approach should shed considerable light on how mathematics begins, functions, and is understood.

"What," Lakoff and Núñez ask, "is the cognitive structure of sophisticated mathematical ideas?" They posit three findings from cognitive science as essential to addressing this question: the first, "the embodied mind," comes from research showing that sensory experiences structure human concepts; the second, "the cognitive unconscious," suggests that most human thought is unconscious—"not repressed in the Freudian sense but simply inaccessible to direct conscious introspection"; and the third, "the conceptual metaphor," names the mechanism by which humans understand abstract concepts through more concrete ones.[26] With these three findings Lakoff and Núñez attempt to explain how abstract, highly formalized mathematical ideas emerge, build on one another, and hold meaning for those who understand them. Absent from all the research on mathematics, they contend, is analysis of the ideas implicit in equations (what Rotman called the "meta-Code"), the ways mathematical ideas are weaved together in mathematical statements, and why the veracity of a mathematical statement comes from those ideas and their interdependence.

One can hardly speak of mathematical ideas, of course, without encountering the Platonic view, and so Lakoff and Núñez do, however briefly. In contrast to Rotman's impassioned challenge, theirs is an agnostic approach: "The question of the existence of a Platonic mathematics cannot be addressed *scientifically*. At best, it can only be a matter of faith, much like faith in a God.... Science alone can neither prove nor disprove the existence of a Platonic mathematics, just as it cannot prove or disprove the existence of a God." Later they clarify that the only mathematics they are interested in is "humanly created and humanly conceptualized mathematics." They then point out that because human mathematics is "conceptualized by human beings using the brain's cognitive mechanisms," questions about the nature of human mathematics are scientific questions, and, if one believes mathematics to exist a priori, the burden of proof lies with that person to prove it scientifically.[27]

Sidestepping Platonic realism in this way has some disadvantages. *Where Mathematics Comes From* is, for example, less alive to Platonic realism's widespread influence and continued hegemony. More important, we hear Rotman shout from the other room, the work, perhaps less reflexive due to lack of engagement, allows certain Platonic tendencies to slip in through the back door. While these critiques bear consideration, their overemphasis might eclipse the great advantage of this approach: their argument makes strategic use of the logic of science—falsifiability—to undermine the claims of Platonic realism even as it legitimizes what is, at bottom, a metaphorical analysis of mathematical discourse.

The most insightful moments in Lakoff and Núñez's treatise, those which set the book apart, emerge through mathematical metaphor analysis. "Many of the confusions, enigmas, and seeming paradoxes of mathematics," they suggest, "arise because conceptual metaphors that are part of mathematics are not recognized as metaphors but are taken as literal." Deliteralizing those metaphors will thus help clarify some of the conflicts associated with mathematics, but, they caution, doing so will not eliminate mathematics' metaphorical content: "Conceptual metaphor ... [enables] us to reason about one kind of thing as if it were another. This means that metaphor is ... a *grounded, inference-preserving cross-domain mapping*—a neural mechanism that allows us to use the inferential structure of one conceptual domain (say, geometry) to reason about another (say, arithmetic)."[28] Sophisticated mathematical ideas, Lakoff and Núñez argue, often weave several conceptual metaphors together; only through metaphoric analysis, then, can we begin to reveal their highly complex conceptual content.

Let us consider an accessible but still interesting example. The Cartesian coordinate plane—a symbolic apparatus that marks for many the beginning of modernism—had a profound impact on mathematics, science, and culture and holds together a blend of metaphors crucial to understanding its conceptual force (see fig. 3). One of the basic conceptual metaphors involved in René Descartes's seventeenth-century invention—what Lakoff and Núñez call the "category is container" metaphor—transforms number into space; that is, before one can conceive of a Cartesian plane one must accept (perhaps unconsciously) the metaphor that maps the category "numbers" onto the spatial representation "number *line*." The number line becomes the container for the category "numbers." To that metaphor one then weds the "geometric figures are objects in space" metaphor that is the basis of the Euclidean plane (and Platonic metaphysics). This metaphorical blend ("The Cartesian Plane blend") achieves a simple but powerful alchemy: it enables one to simultaneously arithmetize space and spatialize algebra. Suddenly mathematicians "can geometrically visualize functions and equations in geometric terms, and also conceptualize geometric curves and figures in algebraic terms."[29] The Cartesian plane thus becomes an efficient means

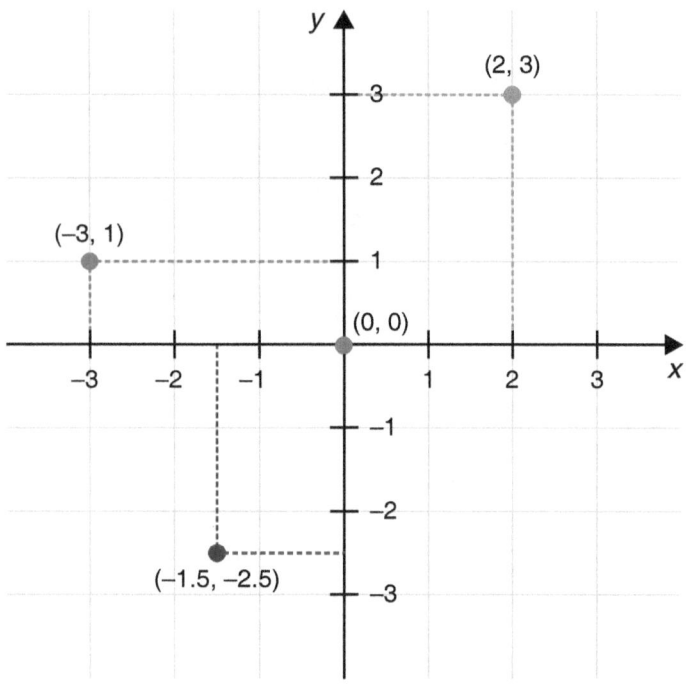

Fig. 3 | Two-dimensional Cartesian coordinate plane

of transformation, for it not only allows functions to be visualized geometrically and geometry to be figured arithmetically but also enables all manner of previously nonmathematical phenomena (terrain, motion, rates of change, weather) to be mathematized—that is, filtered and transformed and mapped into a mathematical space constituted by conceptual metaphors.[30]

Such work marks an exciting moment of possibility: Lakoff and Núñez's research performs a mode of analysis of mathematics whose roots lie in rhetorical studies. Furthermore, this kind of analysis, they argue, is not merely *possible* but *necessary* if one wants to *understand the meaning* of mathematical statements. The implications ripple across many shores. One wonders, building on their provocative claims, how other tropological forms (analogy, metonymy, etc.) function within mathematical discourse; to what extent polysemy (e.g., concepts with multiple meanings) plays a role in historical moments of innovation, crisis, or shift in mathematical practice; and in what ways the arts of mathematical invention might be linked to novel metaphorical thinking.

Provocative as their work is, however, we must temper any enthusiasm with a sober recognition of the book's shortcomings. In spite of their attention to the great variety of conceptual metaphors within mathematics, a dangerous reductivism surfaces in their work. Take, for example, their analysis of infinity. Few concepts in mathematics have had a more profound influence, and few could match the historical variety of formal symbolics through which infinity has been thought and mobilized. Lakoff and Núñez dedicate a whole section of their book to the conceptual metaphors that "govern" the meaning of infinity. Yet their own conceptual methodology, which seems intractably tied to the hierarchical logic of a concrete/abstract binary, forces them to argue in an essentialist fashion that there is one most-originary metaphor of infinity ("The Basic Metaphor of Infinity") on which all other mathematical thought of the infinite is based. If this kind of hierarchical reductivism does not sit well with you, you are in good company: many mathematicians have decried the flattening out of important conceptual diversity that Lakoff and Núñez's work eschews.[31] Importantly, however, mathematicians have not rejected their conceptual metaphor analysis altogether. We can, I think, take the strongest elements of their approach without burdening ourselves with their penchant for origins.

Even after that culling process, though, questions persist. As in Rotman's work, Lakoff and Núñez's mode of analysis affects a microview of mathematics. One wonders over the yet broader issues that their analyses help us to understand but fail to address: What of the mathematization of the human experience gathering momentum in the twenty-first century, tied as it is to a positive feedback loop in which mathematics makes the technologies possible that amplify the need for mathematization? And how does mathematics extend out beyond itself to shape our social-material world? For answers to these questions we must turn to the work of Bruno Latour.

The Translative Approach

Following a tendency in Latour's own work, let's begin with a vignette: King Hiero sits on his throne—troubled by the forces threatening Syracuse—when a letter is delivered from a young man named Archimedes, who (according to Plutarch) made the astonishing claim that "with any given force it was possible to move any given weight." The young Archimedes was so bold, tells Plutarch, that he professed, "If there were another Earth, and he could go to it, he could move this one." Astonished, King Hiero requested a demonstration, leading to the oft-told story of Archimedes raising a "three-masted merchantman of the royal fleet" simply by "setting in motion with his hand a system of compound pulleys." Struck by the potential of his art, King Hiero immediately set Archimedes to work at designing "offensive and defensive engines to be used in every kind of siege warfare."[32]

However incomplete Plutarch's account, it reveals several interesting things about mathematics. Math, according to convention, is the language of abstract thought, the contemplation of the common forms that lie beneath the appearance of things.[33] Math here becomes an escape from and a corrective to the whims and follies of politics, the decisions of which are based on mere opinion (just as Plato opined).[34] But from Plutarch's account (and the many others like it), one can assemble a different view of mathematics, one that does not escape but rather *extends* politics; a practice of thinking that assembles powerful apparatuses of translation through which our social collectives *materially expand*. To explain: from this alternative perspective Archimedes did not simply reveal to King Hiero the secret power of ratios exercised through the technology of compound pulleys; instead, he massively transformed "power relations through the intermediary of the compound pulley"; in doing so, "he also reversed political relations by offering the king a real mechanism for making one man physically stronger than a multitude."[35] Precompound pulley, the sovereign, while representative of his people, was not stronger than his people. Postcompound pulley, the sovereign—allied with a new technology—was suddenly more formidable and thus less indebted to his subjects for his power. How to make sense of this moment of *empowerment*? One could, as Plutarch himself does, tell a story of transcendence, Archimedes becoming the sage, tapping the secrets of nature written in mathematical code. Doing so establishes a clear hierarchy between math and politics: the latter indebted to the former, the former purified of the latter. Latour offers us a different reading.

What is fantastic about mathematics for Latour is how it accomplishes the opposite of what we are told math does. Instead of allowing humans to transcend the political and the social, math in fact actively extends those realms through novel alliances. In the Hiero-Archimedes story we find an emergent alliance between a political form and the compound pulley that materially transforms the social collective. King Hiero's power has expanded not merely through Archimedes's genius but also through a new association between humans and nonhumans (compound pulleys, reengineered siege engines) that Archimedes's mathematical propositions of ratios made possible. But for Latour those mathematical propositions do not reveal an a priori law of nature, and Archimedes did not "discover" said law. Thinking in these traditional metaphysical ways only apotheosizes math while concealing the practice of mathematics as a practice of assemblage, one that, far from separating humans from nonhumans or society from nature, in fact breeds hybrids of humans and nonhumans, societies and natures, materially expanding our social collectives in the process.

Everything sensible seems simple once said, but the difference between this translative understanding and the conventional realist understanding of math is profound. Realists, as we know, make the ontological presumption that mathematical objects exist prior to any human contemplation of them. Ontologically, they are absolute beings that transcend all historical and environmental change; they are, as Russell claimed, that "to which not only the actual world, but every possible world, must conform." Change is thus an illusion, an appearance that conceals the unchanging truths that lie beneath, which only humans (that highest of being) can discover through the forms of pure reason that gave rise to math in the first place.[36] In contrast, a Latourian approach rejects (at least initially) all ontological presumptions and instead begins with practice, seeking to understand not what mathematics "is" but what mathematics does, how it works, and how those who think mathematically practice their art. In doing so he finds that math is not that to which all things must conform but rather a practice of translation that renders what was once incommensurable commensurable.

In this light Archimedes's propositions were powerful because they rendered commensurable what was, prior to the compound pulley, incommensurable: politics and ratios, the one and the many. The consequences were extensive: "Up to that time, the Sovereign represented the masses.... Archimedes procured a different principle of composition for the Leviathan by transforming the relation of political representation into a relation of mechanical proportion. Without geometry and statics, the Sovereign had to reckon with social forces that infinitely overpowered him. But if you add the lever of technology to the play of political representation alone, then you can become stronger than the multitude."[37] The two key phrases here are "principle of composition" and "transforming the relation," for they capture two points essential to understanding mathematics as a translative rhetorical force, one that is reshaping culture in increasingly powerful ways. Point one: mathematical propositions are principles of composition; they are not representations of transcendental truths but actors that enact a recomposition of the existing collective. This means that mathematical propositions have an agency unto themselves that is excessive to human agency. Archimedes was certainly aware of how his mathematics of ratios made commensurable the incommensurability of the large and the small, but one would be hard-pressed to claim he foresaw how those same propositions would recompose the relations of power between sovereign and citizen. These unintended reverberations—or, following Karen Barad, "diffractions"—are traces of the agency of mathematical propositions.[38] Understanding mathematical propositions as actors (or actants) renders any claim that they reflect an a priori reality nonsensical, since they so clearly transform and extend the collective of humans and nonhumans that we call reality. To understand how math works as an embodied symbolic practice that translates, reconfigures, and expands networked relations, then, we must forego the metaphysical logics of representation for the modalities of translation and mediation.[39]

Translation and mediation bring us to point two: like all discursive formations that perdure, mathematical discourse is a powerful system of translation and mediation out of which new hybrids emerge, and those hybrids can, in the right circumstances, "transform the relations" of the networks that compose our world. If we want to understand how mathematical discourse becomes materially manifest, we cannot begin with the presumption of the a priori object, which conceals from view the material practices of inscription, translation, and assemblage that constitute mathematics; instead, "we start from the *vinculum* itself, from passages and relations, not accepting as a starting point any being that does not emerge from this relation that is at once collective, real and discursive."[40] What careful rhetorical study of mathematical practice teaches us (perhaps better than any other discursive form) is that there are not two worlds, one made of symbols and one made of things, but one world of relations and that, while some of those relations certainly existed prior to human thought (relations between oxygen and hydrogen for instance), many others have emerged through the practices of inscription and symbolic action that bind humans and nonhumans together in increasingly novel ways.

How exactly do these practices of inscription give rise to novel relations? Latour offers several examples. Consider, he suggests, how numbers symbolically translate many diverse phenomena into a single commensurable *form*: "*Numbers* are one of the many ways to sum up, to summarize, to totalize ... to bring together elements which are, nevertheless, not there. The phrase '1,456,239 babies' is no more made of crying babies than the word 'dog' is a barking dog. Nevertheless, once tallied in the census, the phrase establishes *some* relations between the demographers' office and the crying babies of the land." The demographers' office, one center of calculation among many, uses numbers to "know" a population. But the power of mathematizing a population does not lie in the numbers themselves; it lies in the concentration of diverse phenomena—age, gender, wealth, religious affiliation—into *one form*: "Were we to follow how the instruments in the laboratories write down the Great Book of Nature in geometrical and mathematical forms," Latour explains, "we might be able to understand why forms take so much precedence. In centres of calculation, you obtain paper forms from totally unrelated realms but with the same shape (the same Cartesian coordinates and the same functions, for instance). This means that *transversal* connections are going to be established in addition to all the *vertical* associations made by the cascade of rewriting."[41]

Numbers, then, are one of the first technologies of mathematization, which is a practice of vertical and transversal rewriting, of translating the world into mathematics—into a formal language that renders commensurable what once was incommensurable. Out of that process many new potential relations emerge. They are *potential relations* precisely because they *do not yet exist*. But once the census data is collected and analyzed, one can begin to link the number of babies in the land with something like fertility rates, which might correlate with pollution or the size of public parks or the quality of schools. Numbers, in short, *manufacture* a commensurability of form that encourages the humans that interact with them to *imagine* novel relations, unconstrained as they are by the radical heterogeneity of everyday life. Those humans can then take those novel relations to those in power (just as Archimedes did), and those novel relations can then lead to the creation of new hybrids, new machines, new institutions that ultimately reconfigure the social-material world.

European exploration of the Pacific in the eighteenth century offers one last example to underscore the profound ways in which mathematical discourse functions as a translative rhetorical force. When Lapérouse first visited Segalien (or Sakhalin) in the Pacific Ocean in 1787, he was weak: he had no knowledge of the land, the navigable straits, or the points of danger; he was dependent on the native population for guidance. Yet, when Europeans returned a decade later, they were stronger—no longer dependent on those same natives. What changed in ten years? The modalities of number and calculation, combined with the Cartesian coordinate system, allowed explorers to extract and mobilize traces through the use of logbooks; those traces, slowly and painstakingly gathered in faraway centers of calculation, enabled the production of navigational maps; those maps facilitated flows of capital, extraction of resources, and exploitation of peoples.

Through numbers and Cartesian coordinates and logbooks, Segalien was subsumed (along with many others) into a broad category of European cartography—in French, Pacifique Oriental; in English, the East or West Pacific—none of which named places as much as they did new relations of power we now call colonialism. And colonialism is in part a name for the desire to control at a distance: "How to act at a distance on unfamiliar events, places and people? Answer: by *somehow* bringing home these events, places, and people." Numbers, Cartesian coordinates, and logbooks combined to form an apparatus that transformed the ragged unknown coastlines of Segalien into the stable forms of East and West Pacific navigational maps. How? The inscriptions in the logbooks, through number and an agreed-on system of coordinates, became "immutable mobiles," which traveled back from Segalien to the centers of calculation in Europe, which then allowed European scientists to create simulacra (navigational maps) of what they increasingly referred to as "the East [or West] Pacific."[42] Those simulacra had a number of material consequences: (1) they allowed Europeans to *simulate* their next exploration before ever leaving shore; (2) as a result, those Europeans became less indebted to the peoples of Segalien (and elsewhere) for their safe passage; (3) the heterogeneity of the peoples and places inhabiting these vast geographies became transformed into the stable forms of the simulacra; (4) the simulacra, in tandem with other cultural forces, slowly translated the subjectivities of myriad peoples into the objectivities of East and West Pacific navigational maps; and (5) those objectivities could then be more easily located, extracted, and sold.

The point of these examples is not to condemn math for reductionism, which would be a massive misunderstanding of how it works, but, on the contrary, to try to understand the unparalleled productivity of the mathematical sciences. Reduction of incommensurability to commensurability, heterogeneity to homogeneity, is only the first moment of mathematization and, if emphasized too much, can eschew the fact that the purpose of such reductionism is not sameness but transformation, differentiation, and extension. Through a reduction to similarity of form, mathematics can reveal potential relations that often materially transform networks. Like an atomic detonation that must first implode before exploding, math first reduces or, better yet, transforms into a common form the elements and constraints of the problem-situation, which occasionally leads to the realization of a novel relation that (again, occasionally) becomes materially manifest in the world. "The sciences multiply new definitions of humans without managing to displace the former ones, reduce them to any homogenous one, or unify them," Latour notes. "They add reality; they do not subtract it."⁴³ Just like Archimedes's compound pulleys, European navigational maps were, in the eighteenth century, novel hybrids that materially expanded the collective of humans and nonhumans that existed at the time.

Rhetoric and Mathematics No Longer Estranged

Latour's arguments, when combined with the work of Rotman, Lakoff, and Núñez, bolster the case for developing links between rhetoric and math. Far from the language of pure reason, through this scholarship we begin to see mathematics as the web that weaves the world of technoscience, as a discourse that rewrites the networks it "represents," and as a vehicle that concentrates power. Mathematics, then, is inherently public and political—all the more so for the efforts that seek to insulate it from these very domains. Mathematics must be understood in light of the networks and relations it renders, the ways it simultaneously concentrates and conceals power, the means by which it translates the objects of its gaze, the productive metaphorical structures it uses, and who we become when we do math. It must, in short, be understood in its myriad rhetorical forms.

Increasing numbers of scholars are already attuned to what we might call the traditional rhetorical forms of mathematical discourse—the ways, that is, that mathematics can advance an ideological agenda, function as an appeal to authority, manipulate audiences, become a means of ethical dissociation, or form a wall designed to obfuscate—a primary concern in critical algorithm studies (see chapter 5) and a strategy Isaac Newton uses to quell deliberation over the Calculus (see chapter 3).⁴⁴ Fewer scholars, however, are attuned to what we might call *the constitutive rhetoric* within mathematics—that is, the role of symbolic action and argumentation in the invention of new mathematics and thus the role of these forms of rhetoric in the evolution and growth of mathematical discourse. And fewer scholars still are attuned to mathematical discourse as a *translative rhetorical force*, which is to say, attuned to the power of mathematical discourse itself and how it translates and transforms fields of power, builds relations of commensurability, and assembles translation machines that refashion the world.

Several implications emerge from thinking of rhetoric as constitutive and math as translative. First, to think along these lines we need to expand our theories of both rhetoric and math beyond the episteme of representation. That episteme, codified in the dialogues of Plato and alive and well in contemporary culture, renders symbolic action, as Rotman noted, a secondary act of representation of "the thing in itself," whether that "thing" is a natural phenomenon or an abstract idea. Inquiry into mathematics dominated by the episteme of representation is thus unlikely to see the liveliness of mathematical discourse: its agential dimensions, its powers of constraint, its terministic screens, or its inventional resources.

Close rhetorical analysis of mathematical statements, however, makes the separation of words and things difficult to sustain. This is simply because mathematical discourse often begins with words but ends with things—that is, with hardened mathematical objects that are themselves hybrids of human and nonhuman agency. Those mathematical objects are often combined into complex propositions that end up *mattering* precisely because, as Karen Barad eloquently put it, "matter is ... not a thing but a doing, a congealing of agency."[45] As we saw with Archimedes's ratios, it takes many human and nonhuman agencies to give rise to a mathematical proposition that *matters*—one that functions as a principle of composition that recomposes the world. This is how mathematics underwrites the worlds of capital and technoscience, not because it is true in some transcendental sense but because it is a powerful form of discursive translation, collapsing distance by mobilizing traces, transforming particularities of difference into stable similarities of form, and, because of that formal coherence, encouraging associations and combinations between phenomena that, left untranslated in the world, have no relation whatsoever. Those relations do not exist a priori, just waiting to be discovered; they are fabricated extensions of the existing collective. Mathematical statements are not merely representations of reality; they are multipliers of reality.

The difference, then, between rhetoric as representative (of some "thing" else) and rhetoric as *constitutive* is substantial: if rhetoric remains the province of representation (as, for example, intentional persuasion), one might reasonably ask how appeals to logos, ethos, or pathos happen within mathematical contexts without ever implicating mathematical concepts themselves, which remain reassuringly arhetorical. As I noted in my first study of the Calculus, within the early scholarship on rhetoric and mathematics, "the argumentative logics that constitute mathematics' productive vocabulary (numbers, geometric figures, axioms, the Cartesian coordinate system) remain unquestioned while the argumentative strategies between mathematicians and the rhetorical influence of mathematics in other discursive fields take center stage."[46] From within the episteme of representation, in other words, it is only the communication of mathematical concepts and not their invention or translative force that would be of proper interest to rhetorical scholars.

Constitutive rhetoric is a conceptual orientation from which to see mathematics anew. Mathematical objects are typically conceived as existing independent of human cognition. Mathematical symbols are, accordingly, more or less adequate representations of ideal mathematical objects, and, as a result, mathematical discourse is always subordinate to genuine mathematical thought and the ideal objects that thought contemplates. As such, any role that rhetoric might have regarding mathematics is largely parasitic (à la Plato), at best helping spread mathematical truths and at worst actively obscuring them. This is because, for the mathematical realist, symbolic action is an imperfect medium for the *representation* of a priori mathematical objects. A shift to rhetoric as constitutive, however, challenges this realist paradigm and the metaphysics that traffic with it. Instead of considering mathematical statements as better or worse representations of preexisting mathematical objects, a constitutive approach encourages scholars to imagine symbols, inscriptions, and arguments as the material out of which mathematical concepts emerge, coalesce, and are (if successfully articulated) integrated into the existing mathematical lore. Within this context to say math is rhetorical is not to say math is persuasive or manipulative; it is to say that math is a species of symbolic action that enables relations to be taken up, translated, reconfigured, and occasionally transformed into something new. Thus rhetoric plays a significant role not just in the communication of math but in the very practices of innovation and invention within mathematics. Rhetoric, as embodied material practices of inscription and argumentation, is not an enemy or an obstacle to mathematics but rather an engine of its evolution. From a constitutive perspective what is most interesting about math is not the a priori reality it reveals but how mathematical propositions and apparatuses—such as the golden ratio or the Cartesian coordinate system—actively constitute realities as they emerge and interact with our social-material worlds.[47]

Mathematics thus never "purges" itself of rhetoric, understood properly as constitutive embodied inscription. As Latour noted in *Science in Action*, we see again here that mathematics does the opposite of what we are often told it does. For as a mathematical statement evolves from informal conjecture to formal proof, more not less symbolic resources weave around and into it—more relations are established and more networks built, slowly making the mathematical statement harder, more semiotically dense, more resistant to challenge, and more real (a process imaginary numbers put on spectacular display in chapter 4). Once hardened through the rigors of mathematical argument, a mathematical statement can become, on occasion, a principle of composition, able to introduce novel relations, give birth to new hybrids, and thereby materially expand the network of relations that constitute our social-material world.

Perhaps a visual diagram would help us "see" this translative process and the relations therein more clearly. In figure 4 one begins with an intra-action between the social-material world and embodied inscriptive practices (intra-action because inscriptive practices are always of and in the social-material world). This intra-action gives rise to a process of selective and deflective translation and formalization—all of which enables commensurable forms of in-*form*-ation to emerge. Out of that information demarcated problem-situations take shape. Informal mathematics names processes and practices mathematicians use to understand and test the relations and constraints of a particular problem-situation, which includes (among others) diagramming, concept-stretching, cross-domain mapping, and running thought-experiments. Much of that effort ends in productive conjecture failure (because one often learns something about the problem-situation even in failure), but on rare occasions one happens upon a potentially viable novel conjecture, a conjecture slowly hardened by mathematical communities into a new principle of composition, the uptake and application of which can give rise to new hybrids, expanding and transforming the social-material world in the process.

Much more than a means for discovering truth, then, mathematics emerges from this perspective as an increasingly powerful translative force. Mathematics is not a "thing" to be defined (by the *Oxford English Dictionary* or anyone else); it is a dynamic practice of inscribing, imagining, translating, reasoning, and arguing in virtual worlds often structured by formalisms (number, geometry, topology) that enable patterns and relations to emerge. Note in this description of math the arts of rhetoric (inscription, argumentation, relational formation) are immediately incorporated into our understanding of mathematics, not as an "abstract science" but as an embodied practice of inscriptive, innovative imagining, thinking, and translating. This, I submit along with Lakoff and Núñez, is the only math we know—an embodied practice of discursive and nondiscursive translation via an infinite (because always unfinished) variety of formalisms that simultaneously reveal and constitute patterns and relations in a singular symbolic-material world. There are not two worlds—as both realists and constructivists

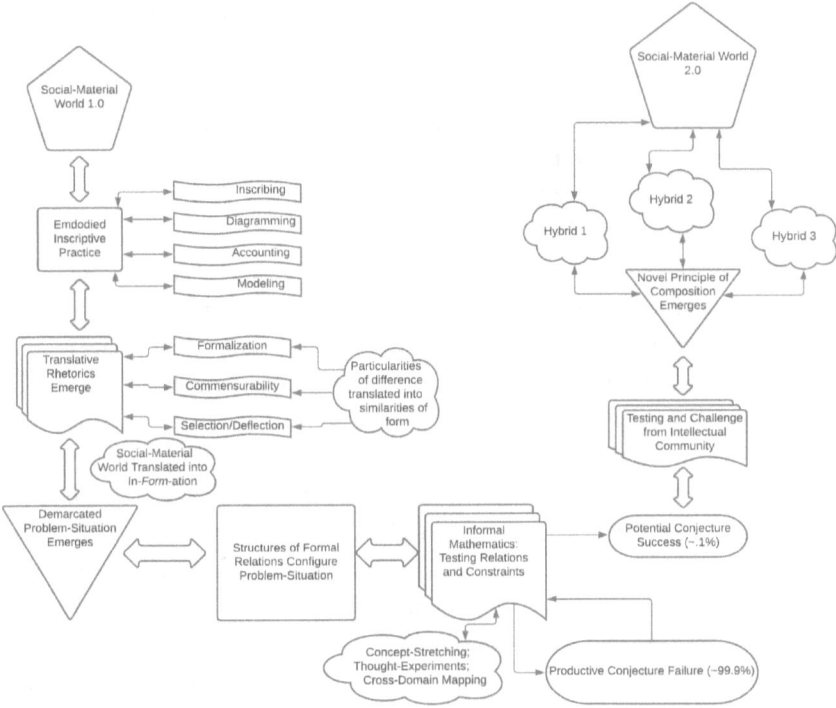

Fig. 4 | Diagrammatic model of translative rhetoric

often presume—but only one world of symbolic-material relations. That's all reality is—a complex network of relations, some easily transformed (what we might call "soft" relations) and some incredibly difficult to transform (which we can call "hard" relations); some much older than humans (e.g., relations between sun and planets) and some newly born (relations between humans and the internet). Transforming soft relations typically requires less energy (effort, understanding, techne, resources; etc.), while transforming hard relations often requires more (think, for instance, of the difference between separating liquids and splitting atoms).

These new senses of rhetoric and mathematics challenge us to rethink how we see their relations and potential interactions. Rhetoric as constitutive, for instance, implies the study of the realm "symbolic action," of which "persuasion" is a specific species. In this realm there are many symbolics (countless species, countless forms), all of which simultaneously represent, constitute, and transform the social-material world. Rhetorical scholars study this dynamic ecology of symbolic action, how particular symbolics take up existing relational networks, in one moment seeking to harden those relations and in another seeking to sever them; how they translate (or try to translate) the world into logics of their own making; and how they can, on occasion, introduce new relations and thereby profoundly transform and expand the social-material world (see fig. 4). Mathematics names a collective of symbolics (for there are many mathematics) that often share some basic formalisms (e.g., numeric, geometric, algebraic), which enable it as a species of symbolic action to represent, constitute, and transform relations more powerfully than most other symbolics. Humans create and interact with these symbolics, but they do not control them. Often "our" symbolics of communication shape and constrain us just as much as we shape and constrain them.[48] This fact, which has always been the case but is easier to see today, with our rapidly proliferating modalities of communication, forces us to go beyond anthropocentric definitions of symbolic action that imagine humans as the central agent and symbolic apparatuses (whether discursive or material) as mere tools. As several decades of scholarship has now shown, such humanistic paradigms have an increasingly difficult time accounting for symbolic action in the twenty-first century, a place where trees communicate, animals speak and reason, algorithms make decisions, and artificial intelligence increasingly shapes our lives.[49]

Rhetoric as constitutive and mathematics as translative is my way of attempting to escape the gravity of anthropocentric humanism. Math as translative also reflects my understanding of mathematics as a dynamic practice of innovative imagining and reimagining, one where ethical argument (e.g., an honest exchange of ideas) is essential and often constitutive of mathematical concepts, and where that argumentation seeks not merely its own self-fulfillment but rather to form or harden provocative conjectures with the potential to introduce something radically new. The notion of math as translative rhetoric also seeks to describe the practice of mathematics as a practice of translation and transformation, the emphasis of which helps explain how mathematical discourse has become so influential in the twenty-first century, a point the chapters of this book seek to demonstrate throughout.

Finally, I must emphasize again that to describe math as translative rhetoric is not to say math is manipulative or persuasive (though it can be bent to such purposes), but rather that mathematical discourse is a powerful form of constitutive symbolic action. The statements emerging from the symbolic-material practice we call mathematics can, on occasion, function as *principles of composition*, which means they do not merely represent a preexisting reality but, far more interestingly, they can and sometimes do actively recompose and expand reality (that is, the collective of relations that we call reality). Tracing those translations, innovations, and recompositions will, I hope, help us come to grips with the ways public discourse and public culture are rapidly and irreversibly transforming in the twenty-first century.

Implications: Mathematics as Translative Rhetorical Force

By way of summary, over the past two chapters we have traced how Plato positioned rhetoric and mathematics as oppositional forces. We have seen how his philosophy of mathematics reduced it to a vehicle for discovery of *aletheia*, concealing in the process mathematical practice as an embodied discursive practice of assemblage, translation, and transformation. And we have seen how Plato's representationalist theory of language transforms the careful, often painstaking, work of mathematical diagramming and interactive inscription into inevitably flawed imitations of the real mathematical objects to which they supposedly gesture. As a result, and due to the widespread influence of Platonic realism, the rhetorical force of mathematical discourse (which is to say, its generative, translative capacities) has largely been ignored.

Only recently does one see scholarship directly challenging the estrangement between rhetoric and mathematics, implying that conventional definitions of both rhetoric and math increasingly fail to account for their dynamic interaction. Rhetoric quarantined as manipulation or persuasion will no longer do as a paradigm for inquiry and understanding of contemporary symbolic action. Mathematics as abstract contemplation of ideal objects will no longer do as a paradigm for inquiry and understanding of mathematical practice, much less how math is rapidly translating and rewriting culture. In fact, Platonic realism, while perhaps intoxicating to the human ego, has never been able to account for mathematical growth.[50] What I hope to show in the remainder of this book is that a rhetorical approach to mathematics can offer much insight into both the evolution of mathematics and how that evolution has led to our contemporary moment—one in which mathematical discourse is an increasingly powerful translative rhetorical force.

3

Transgressing the Limit | Invention, the Calculus, and the Rhetorical Force of the Infinitesimal

How does mathematics grow? How has it evolved over time? Addressing such questions calls not for study of the finished products of mathematical labor, polished and refined such that their messy rhetorical processes of invention, inscription, argumentation, and translation are but distant marks on the horizon, if not erased altogether. No, for an understanding of the evolution of mathematical discourse and how that evolution has accelerated the reshaping of our social-material world, we must attend to mathematics-in-the-making. We must examine math in moments of radical upheaval and transformation such that we might begin to glimpse its evolution in process. The end of the seventeenth century was such a time for mathematics (and, truth be told, for Western intellectual culture). It was a time of mathematical revolution whose magnitude continues to reverberate to this day.

Until the 1660s the standard-bearer of truth in mathematics was Euclidean geometry, and the near-universal philosophy among mathematicians was one form or another of Platonic realism (the titles "formalist" and "foundationalist" were popular at the time). The purpose of math, accordingly, was thought to be the discovery of absolute truth, and mathematical discourse was considered an imperfect vehicle for the representation and communication of said truth. All of that, however—the security of the Euclidean foundation, the idea of mathematical discourse as referential to *real* mathematical objects, the notion of rigorous math as an infallible logic of truth—was about to be upended with the emergence of "the Calculus" and the use of infinitesimals in Sir Isaac Newton's *Methods of Fluxions and Infinite Series* and Gottfried Wilhelm Leibniz's *Acta eruditorum*.[1] In this chapter I show that, in using the infinitesimal, Newton and Leibniz introduced an extraordinarily productive mathematical concept into seventeenth-century mathematics that is *only rhetorically substantial*. The infinitesimal, in other words, is an exemplar of the highly productive work of constitutive rhetoric in the invention, growth, and transformation of mathematical discourse. In addition, with the rise of the Calculus and infinitesimals, we also see an evolution in mathematical discourse beyond the logics of representation, a phenomenon that, I argue, is highly correlated with the increasing translative power of mathematics in the social-material world.

The phrase "a Calculus" refers simply to a system of mathematics. That one particular achievement in mathematics should take this generic term and make it its own speaks to that achievement's importance. Before Newton and Leibniz invented integral and differential Calculus, the term *calculus* was generic. Since then, Newton's and Leibniz's Calculus has become popularly regarded as "the Calculus." The identification of the term *Calculus* with Newton's and Leibniz's methods is easily understood in light of what it meant for mathematics and the physical sciences, but its rhetorical character remains enigmatic.

Newton's and Leibniz's Calculus allowed for the study of velocity, acceleration, and, generally, the movement of objects through space. At the most basic geometrical level, the Calculus breaks the area under a curved line into an infinite number of rectangles, transforming the curve into an infinite number of straight lines and estimating the area under that curve by summing the infinitesimal rectangles (thus the original name, infinitesimal Calculus). In practice, of course, summing an infinite number of rectangles is impossible. The key is to understand the logic. As the number of rectangles increases (as in fig. 5) the error in approximating the area under a curvature converges toward zero. If one sums enough rectangles, one is eventually left with what Newton and Leibniz considered an infinitesimal error, something that could be profitably neglected.[2] A short time after the original method emerged, Newton and Leibniz found that, in summing the rectangles and estimating the area under a curve, one was also estimating the distance an object traveled in motion; they then extrapolated methods to approximate and predict rates of change, including velocity and acceleration. In this way the logic of the infinitesimal became instrumental to solving pragmatic mathematical and scientific problems.[3]

Regarding the importance of these findings, Howard Eves writes, "[the] calculus marks a watershed ... in the history of mathematics." Eves emphasizes its importance, calling mathematics before the Calculus static, "the still-picture stage of photography."[4] Extending Eves's metaphor, Newton's and Leibniz's Calculus was the moving-picture stage for mathematicians and scientists; it allowed mathematicians, physicists, astronomers, and countless others to move

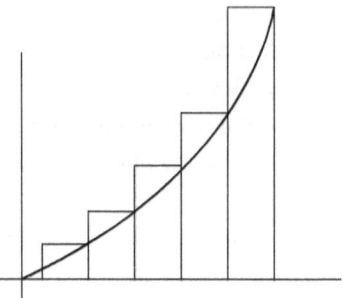 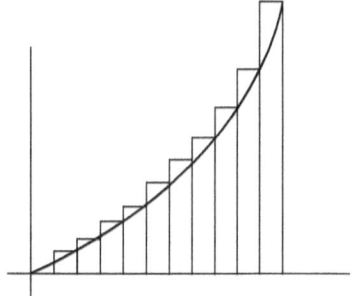

Fig. 5 | This example demonstrates that, when the number of rectangles used to approximate the area under a curve is doubled, the error is reduced. In addition, if one were to connect straight lines from rectangle to rectangle in a consistent manner, one would notice that, as the number of rectangles increase, the continuous shape of the original curve is better approximated.

beyond the study of finite spaces and fixed points and into the world of movement, change, and the infinite.

The Calculus initiated changes in rhetorical practice (e.g., in language use and the modes of appeal available) as great as the changes it initiated in mathematics and physics.[5] Martin Heidegger argued that Newton's and Leibniz's "axiom begins with *corpus omne*, 'every body.' That means that the distinction between earthly and celestial bodies has become obsolete. The universe is no longer divided into two well-separated realms. All natural bodies are essentially of the same kind."[6] The Calculus collapsed the distinction between arbitrary points both in studies of mathematics and in ways of thinking; it discarded the traditional distinction between earthly and celestial in terms of motion and thought and expressed thinking of the world in terms of a continuum, thus, providing an alternative to the Aristotelian concept of distinct antithetical entities. As Gregory Desilet writes, "This new understanding of motion provided a model for every opposition, polarity, and dialectic: the contrast can be seen not as an *agon* between two discrete and separate entities or qualities but as a continuum between two extremes."[7] It can be reasonably argued, then, that the Calculus demonstrated a major moment of change regarding the way humans perceived, thought about, and talked about their world, but the Calculus did not operate in a vacuum. The new episteme that the Calculus brought forth surfaced among a network of discourses; it had to be communicated, discussed, and clarified before it could become a dominant mathematical form. In this process—in the rise of the Calculus and the study of the discourses that conditioned its emergence—a kind of mathematical rhetoric finds voice.

This chapter seeks to attune our ears to the rhetoric operating within and around the field of mathematics. Focusing on a period when the ideas concerning (and methods of doing) mathematics were in flux, this chapter ultimately exposes rhetoric's role not just in the external communication of mathematics but also in the invention of a whole system of new mathematical concepts that have governed scientific discourse since. Rather than merely supplemental or a cautionary tale of sorts, rhetoric here figures as the "material" on which a new form of mathematics becomes possible.[8]

In what follows I explore a foundational concept of Newton's and Leibniz's Calculus: the infinitesimal. My analysis examines their treatment of infinitesimals, showing that the concept was the crux of their new Calculus and that employing them necessarily depended on rhetoric.[9] I then turn to Newton's and Leibniz's rhetorical responses to the issues broached by their infinitesimals. Thereafter, I consider the impact of Newton's and Leibniz's "infinitesimal rhetoric" on subsequent discourse, and I conclude with a discussion of the implications of this study for understanding the translative rhetorical force of mathematical discourse.

The Senses of Rhetoric in Mathematics

As we already know, an increasing cohort of scholars have begun to investigate the rhetorical qualities of mathematics. The great majority of this research engages with what we described in chapter 2 as math's conventional rhetorical forms. Philip Davis's and Reuben Hersh's essay "Rhetoric and Mathematics," for example, is a pioneering effort to disclose the rhetorical features of mathematics, to show, in their words, "that mathematics is not really the antithesis of rhetoric, but rather that rhetoric may sometimes be mathematical, and that mathematics may sometimes be rhetorical." In their essay one gains the double sense in which Davis and Hersh mean to discuss mathematics as rhetorical: in the first sense they consider how mathematics is *used* rhetorically in other fields (in economics, for example). In the second sense Davis and Hersh maintain that there are "rhetorical modes of argument and persuasion" at play in the *doing* of mathematics.[10] Here they are interested in showing the nonfoundational nature of mathematical proof.

In both cases Davis and Hersh deploy a rhetoric highly Aristotelian in flavor, one bent on disclosing the available means of persuasion surrounding different forms of mathematical discourse.

Other scholars less explicitly invested in the rhetoric of mathematics have considered everything from the influence of numbers on persuasive appeals to the importance of persuasion for mathematics instruction.[11] In each case, however, rhetoric emerges only as supplemental. Rhetoric is not the "stuff" of mathematics but operates only as the handmaiden to mathematical findings.[12] I do not condemn such research (in fact, it can be highly productive) but merely wish to highlight the direction of its focus and suggest that there are other equally important rhetorical modalities at play in mathematical practice. Indeed, with the *invention* of novel mathematical symbols, a different kind of rhetorical process takes shape, one that remains largely unexamined.

The central claim of this chapter is that Newton's and Leibniz's idea of the infinitesimal is a rhetorically constituted concept that influenced disciplinary practices within and outside the field of mathematics. There are at least two layers of rhetoric circulating around the concept of the infinitesimal. One layer is constitutive in nature, marking the discursive formations that give shape and "substance" to the infinitesimal. Rhetoric in this sense features the arguments that allow for certain ideas to rise up and become dominant explanatory concepts. The infinitesimal has no recourse to empirical verification; there exists no means by which to confirm or deny its existence. Rather, the infinitesimal finds its constitution in the forms of rationality and intuitive logic that saturated late seventeenth-century European mathematics. The infinitesimal's rhetorical substance makes it unique because it pressed beyond the boundaries of seventeenth-century mathematical and scientific practice *simultaneously*.[13] Unlike most mathematical and scientific discoveries until that time, there was no neat mathematical proof with which to fit the infinitesimal into previous mathematics. In fact, Euclid's *Elements* explicitly excluded the consideration of infinitesimals:

3. A *ratio* is a sort of relation in respect of size between two magnitudes of the same kind.
4. Magnitudes are said to *have a ratio* to one another which are capable, when multiplied, of exceeding one another.
5. Magnitudes are said to *be in the same ratio*, the first to the second and the third to the fourth, when, if any equimultiples whatever be taken of the first and third, and any equimultiples whatever of the second and fourth, the

former equimultiples alike exceed, are alike equal to, or alike fall short of, the latter equimultiples respectively taken in corresponding order.

6. Let magnitudes which have the same ratio be called *proportional*.[14]

These definitions barred infinitesimals by requiring that in a ratio a:b continued multiplication will make one exceed the other, thus eliminating the possibility of dividing by zero or its geometric equivalent (infinitesimals).[15] Thus, at the same time that infinitesimals defied empirical testing, they also broke with accepted mathematical practice. Argument was the sole champion of the infinitesimal.

The plane of constitutive rhetoric out of which the concept of the infinitesimal emerged gave rise to a second "layer" of rhetoric, one more conventionally recognized in academic circles. With the infinitesimal's special rhetorical status arose a situational rhetoric characterized primarily by criticism, debate, and response. Indeed, the debate between Newton, Leibniz, and their critics emerged from the infinitesimal's violation of the accepted limits for mathematical and scientific practice. The transgressive logic of the infinitesimal placed the rules that governed seventeenth-century scientific and mathematical practice in question, calling for a critical response. Parallel to Kenneth Burke's understanding of the relationship between identification and persuasion, then, I view the constitutive rhetoric of the infinitesimal as the condition for the possibility of its situational rhetoric.[16]

As with other scholars who have worked to illuminate the relationship between rhetoric and mathematics, the goal of this chapter is to uncover the rhetorical dimensions of mathematics that allow for certain methods to arise and become accepted as dominant mathematical models. By focusing explicitly on the invention of a novel mathematical concept, I show the role of rhetoric in the constitution and dissemination of the Calculus. A purely intentional, agent-centered model of rhetoric (like the one deployed by Davis and Hersh), however, is insufficient to the task of understanding the rhetorical force of the infinitesimal. There is an excess to seventeenth-century mathematical and scientific intention in the infinitesimal that raised debate in the first place. To analyze that excess one must go beyond an intentional (and highly anthropocentric) notion of rhetoric. This is not to suggest agency and intention play no role in my analysis (they are both fundamental); rather, I want to combine an agent-centered approach with a discursive one to show that the infinitesimal itself, far beyond the intentions of Newton or Leibniz, required a rhetorical response.

To complicate intentionality in my analysis, I discuss Newton's and Leibniz's responses to certain criticisms as responses not so much to criticisms by individuals as to a critique generated by a whole discourse of thought (an episteme).[17] Newton and Leibniz were not responding so much to George Berkeley or Bernard Nieuwentijdt (who was the first to criticize the use of infinitesimals) as they were to the rules that animated seventeenth-century mathematics and science.[18] These rules find expression through such luminaries as Nieuwentijdt and Berkeley, but, when they speak, a whole discourse of power speaks. This phenomenon is less the case with Newton's and Leibniz's rhetoric constituting the infinitesimal because that was a novel concept and, as such, did not have immediate recourse to an already established set of discursive rules.

In the last analysis this chapter seeks to understand not only the rhetorical debates between Newton, Leibniz, and their contemporaries but also the rhetorical force of the infinitesimal—that is, the variety of arguments that constituted the concept of the infinitesimal and opened up a radically nonrepresentational realm of mathematical practice. As counterintuitive as it may sound, the infinitesimal had its own translative rhetorical force, a force independent of any author, and seventeenth-century intellectuals found themselves forced to respond. In those responses to the presence of the infinitesimal one can detect the dominant discursive formations governing seventeenth-century scientific practice. Throughout this chapter, then, I discuss Newton, Leibniz, and other seventeenth-century intellectuals as actors responding to one another *and* to the infinitesimal's translative rhetorical force.[19]

Infinitesimals and the Calculus

Newton's and Leibniz's initial accounts of the Calculus were exasperating; they were bare bones, composed mostly of demonstrations and definitions free of proof or explanation, and they did more to scare off those interested in the new method than to entice them. Although in later articles they attempted fuller explanations, one of the most curious features of their whole effort was that none of their contemporaries knew with satisfying clarity just what they were doing or how they were doing it.

The ambiguity of the Calculus arose in large part out of their use of infinitesimals. The idea of applying the infinitesimal in mathematics was radical enough (see Euclid's axioms), but Newton and Leibniz pushed their utilization by creating a general method employing infinitesimal quantities. Part of the difficulty was in thinking of something as evanescent while treating it quantitatively. From all logical standpoints this seems a contradiction—to be able to add up "nothings" and get something.[20] The question, then, is not how the Calculus worked, but how Newton and Leibniz conjured up such a powerful method by employing the abstract idea of infinitesimals.

In their work Newton and Leibniz "relied on the concept of an infinitesimal quantity, that is, a quantity smaller than any positive quantity and yet larger than zero. But, despite their centrality to the Calculus, such infinitesimals ... proved to be confusing and frequently contradictory."[21] With their method Newton and Leibniz were attempting to model mathematically the continuous movement they observed in nature. Unfortunately, mathematics was numerically based, and numbers, no matter how close together, are iterative; they are points one can connect with a line, but they always require connection. What Newton and Leibniz sought was a concept that would allow mathematics to model continuous motion. Ingeniously, they both realized that if one could *imagine* infinitesimal quantities, quantities so slight as to be "smaller than any positive quantity and yet larger than zero," then these quantities could model continuous motion; that is, if one were to divide the area under a curve into an infinite number of sections, one could approximate the velocity function of that curve with negligible error, or, for Leibniz, an evanescent error, one with a mysterious "vanishing" quality.

These infinitesimals, then, were ghosts of a sort. At the same instant that they vanish, they become the most helpful, allowing for such a precise mathematical model of continuity that the error is ephemeral, evanescent, fleeting, as it were; but Newton's and Leibniz's use of infinitesimals was hardly straightforward. When pushed on the subject, they gave a variety of accounts, claiming that their infinitesimals had a polysemous character, able to be summed, "ratioed," and ignored all in the same instant. The concept of infinitesimals appeared to be indeterminate, and, as Michael Bishop puts it, "The possibility of expressing a partially indeterminate concept is an important (and neglected) element of scientific practice. Indeterminate concepts allow scientists considerable flexibility to respond in the future to very different kinds of criticisms that might come from very different quarters."[22]

Although the indeterminate definition of infinitesimals gave Newton and Leibniz more room to negotiate criticism, it also invited attacks. One of the central issues infinitesimals raised was clarity.[23] The standard approach of scientists and mathematicians—a legacy of Platonic realism—was to behave as if there existed a "pure" cognitive language used by a specific community for clear communicative transactions. Inevitably, however, a scientist will have a thought that the current symbolic system does not express. In these cases scientists must create new metaphors to represent their ideas. Often these metaphors are met with confusion and criticism and eventually are either accepted and integrated into the language system or rejected as unnecessary.[24]

Newton and Leibniz created a number of metaphors to account for their notion of infinitesimals, and the sounds of confusion and criticism they heard were the sounds of their scientific community slowly accepting and integrating the essential concept of the infinitesimal into their framework. The lack of a clear and universally accepted definition, however, raised several issues regarding mathematical reasoning and metaphysics. Because the mathematical community was bent on definition, consistency, and rigorous proof according to the axioms of Euclidean geometry, the indeterminate nature of infinitesimals was threatening. Using infinitesimals challenged the integrity of mathematical foundationalism—that is, of mathematical symbols emerging from irrefutable first principles. Infinitesimals brought inconsistencies, and, as much as mathematicians take "intense delight in mathematical theorems that are conceptually outrageous, unbelievable,...'almost false,'" at the same time they have great anxiety when the limits of their language are transgressed, leading to contradiction or paradox. Moreover, infinitesimals "belonged to the realm of theology and faith and not that of mathematics, which of necessity occurs within the secular boundaries of finite human reason."[25]

The constitutive rhetoric of the infinitesimal, then, generated two primary criticisms that Newton and Leibniz would have to address. The first highlighted the apparent paradox of "infinitesimals in degree"—that is, in the way the Calculus deployed greater and lesser infinitesimals. Critics of the Calculus asked how one could define two things as infinitesimal and simultaneously make claims as to their relationship in finite, quantitative terms.[26] Such a claim explicitly contradicted the rules of Euclidean geometry laid out many centuries prior. The second and perhaps more powerful critique struck the language of ratios Newton and Leibniz deployed later in their careers. Critics argued that ratios implied a magnitude that infinitesimals lack by definition. How can one talk of dividing what were referred to (by Newton at least) as indivisibles outside the realm of paradox and absurdity?[27] These were the questions Newton and Leibniz were asked. By virtue of the infinitesimal, Newton and Leibniz found themselves in the difficult position of defending something that transgressed the boundaries of seventeenth-century thought. The vagueness of infinitesimals, the expansion of mathematical language, the resulting inconsistencies, the metaphysical and ontological questions—Newton and Leibniz would have to negotiate all of these using the persuasive resources of language.

Newton's and Leibniz's Rhetoric Regarding Infinitesimals

Owing in part to their personalities and in part to their philosophies, Newton and Leibniz responded differently to questions and criticisms regarding their use of infinitesimal quantities. Leibniz, a more gregarious, didactic spirit, entered into the controversy, spending page after page working out the philosophical implications of infinitesimals. His engagement with the issues infinitesimals raised contrasted sharply with Newton's reticence, perhaps caused by his "morbid fear of opposition."[28] In any case Newton did not readily engage the criticisms of his contemporaries.[29] The archive of his rhetorical treatment of infinitesimals is thus limited. At the same time, Leibniz's rhetorical engagement with the issues raised by infinitesimals reaches nearly unmanageable proportions. Nevertheless, both Newton and Leibniz dealt with infinitely small quantities by appealing to the law of continuity. Although Newton had some philosophical difficulties with the continuum and referred often to the less overtly metaphysical notion of continuous motion, when pushed, "he too, in his prime and ultimate ratio, was implicitly invoking the law of continuity."[30] My present task is to present Newton's and Leibniz's rhetoric concerning the infinitely small and to explain how they went about responding to the exigencies infinitesimals raised.

Near the end of his career Newton, the meticulous scientist, did not wish to speak of infinitesimals in his work because for him they indicated a certain recklessness, a neglect, no matter how small.[31] Newton considered himself a scientist, and for him mathematics was only useful inasmuch as it served the interests of his scientific endeavors. Newton believed "science was composed of laws stating the mathematical behavior of nature solely—laws clearly deducible from phenomena and exactly verifiable in phenomena—everything further is to be swept out of science, which thus becomes a body of absolutely certain truth about the doings of the physical world."[32] From this philosophical stance (which squares with the dominant mode of scientific thought in the seventeenth century), one can easily see that the concept of the infinitesimal would present problems. No matter how one thinks it, rephrases it, or deploys it, the infinitesimal requires neglect, thereby threatening the foundationalist stance Newton occupied.[33] Equally important, the apparent neglect of the infinitesimal opened the door for critics to challenge Newton's method and his work, something Newton loathed.[34] In the beginning, however, when Newton's Calculus was still taking shape, he was less conservative about his use of infinitesimals (and less aware of their implications, no doubt). In fact, Augustus De Morgan showed in one of the first historical articles on infinitesimals that, until 1704, Newton's work freely employed infinitely small quantities: "So far as algebraical Calculus was concerned, Newton himself used infinitely small quantities" regularly. In 1704 Newton published *Quadratura curvarum*, attempting to avoid the use of infinitesimals. As De Morgan asserted, "In 1704 Newton in the *Quadratura curvarum* renounced and abjured the infinitely small quantity; but he did it in a manner which would lead one to suppose that he had never held it."[35] The question of interest here is what pressures, what realizations, what discourses allowed Newton to express positions so obviously in conflict with one another?

To begin to address this question, let's compare Newton's early discourse regarding infinitesimals with his later statements. In the first edition of his *Principia*, Newton stated, "But take care not to look upon finite particles as such. Moments, as soon as they are of finite magnitude, cease to be moments. To be given finite bounds is in some measure contrary to their continuous increase or decrease. We are to conceive them as the just nascent principles of finite magnitudes." In the second edition Newton stated, "But take care not to look upon finite particles as such. Finite particles are not moments, but the very quantities generated by the moments. We are to conceive them as the just nascent principles of finite magnitudes."[36] At first glance these two statements do not appear dramatically different. They are both phrased awkwardly, and they both refer to "moments" or increments of an infinitesimal magnitude. In the first edition, however, it is clear that Newton makes moments and infinitely small quantities equivalent. In the second he makes no such equivalence, stating only that finite quantities are not moments, but that moments can, if summed, generate a finite figure. As soon as they reach that finite posture, however, they cease to be moments. It may seem that I am splitting hairs over Newton's method of expressing a novel and complex concept, but this was only Newton's first rhetorical move *away* from the use of infinitesimals (or what he called "moments").

In 1704 Newton's *Quadratura curvarum* introduced the foundations of his Calculus, avoiding "all use of infinitely small constants."[37] Rather than considering moments as consisting of infinitely small magnitudes, he described them in terms of continuous motion. To eliminate the use of infinitesimals in his work, he adopted "the doctrine of prime and ultimate ratios of finite differences."[38] Described in terms of motion, these prime and ultimate ratios were equivalent to prime and ultimate velocities. For Newton ultimate velocity meant "that with which the body is moved, neither before it arrives at its last place, when the motion ceases, nor after; but at the very instant when it arrives." Similarly, his ultimate ratios of finite differences were to be understood as "the ratio of the quantities, not before they vanish, nor after, but that with which they vanish."[39] Thus, he had rearranged his use of infinitesimals, emphasizing their ratios rather than using them explicitly in his work. This move effectively placed a "layer" of mathematics between Newton's Calculus and his use of infinitesimals.

Evading the focus on infinitesimals and their use, Newton organized his language so the emphasis was placed on the ratios of infinitesimals rather than on infinitesimals themselves. He codified this position in the third and final edition of the *Principia*:

> Because the hypothesis of indivisibles seems somewhat harsh, and therefore that method is reckoned less geometrical, I chose rather to reduce the demonstrations of the following propositions to the first and last sums and ratios of nascent and evanescent quantities, that is, to the limits of those sums and ratios. . . . Therefore if hereafter I should happen to consider quantities as made up of particles, or should use little curved lines for right ones, I would not be understood to mean indivisibles, but evanescent divisible quantities; not the ratios of determinate parts, but always the limits of sums and ratios.

This was Newton's final stance on the application of infinitesimals in his work, and one might note the conspicuous absence of the term *moment*, Newton's word for representing the logic of the infinitesimal. He had effectively moved from using infinitely small quantities early in his career to emphasizing the "ultimate proportion of evanescent quantities" by the end.[40] Why the change? Was Newton responding to criticism? What elements, what discourses, allowed for the evolution?

It is reasonable to conclude that in Newton's initial treatment of the Calculus he was either ignorant of or inattentive to the complications that came with infinitesimals.[41] Comparing the first edition of his *Principia* with the third demonstrates a clear transformation in his thinking about and use of infinitesimals.

Newton did not readily engage in the debates over and criticisms of his work, so it is difficult to determine with certainty the forces at play. Nevertheless, the revisions in his texts dealing with infinitesimals reveal the impact of criticism.[42] In the final edition of the *Principia*, we find one of Newton's few acknowledgments of criticisms concerning infinitesimals: "Perhaps it may be objected, that there is no ultimate proportion of evanescent quantities; because the proportion, before the quantities have vanished, is not the ultimate, and when they are vanished, is none."[43] This sentence is the most we get from Newton, but it offers insight into the criticisms he was negotiating.[44]

There was an inherent difficulty in talking about something from a foundationalist perspective when that concept was unrepresentable and defied verification by experiment. Newton's contemporaries thus claimed his infinitesimals unscientific and unempirical; that is, infinitesimals were purely hypothetical notions with no basis in physical reality.[45] This criticism struck Newton hard; he believed the "best and safest method of philosophizing . . . [was] to investigate the properties of things and establish them by experiment, and then to seek hypotheses to explain them."[46] Infinitesimals, however, were outside the empirical realm, and the only method of proof Newton could use for their existence was *reasoned argument*. This troubled Newton, a pure foundationalist, throughout his career, causing him to revise his use of infinitesimals several times.

Newton's initial response, then, was to evade the use of infinitesimals in his work. By the end of the seventeenth century, he found that it was not the infinitesimals that were of importance, but the ratios of indivisible quantities that allowed for the Calculus to describe continuous motion. The result was a description of the Calculus in the language of prime and ultimate proportions. Focusing on ratios allowed for a more rigorous mathematical proof, but that mathematical rigor did not amount to a conceptual improvement; it simply inserted a thicker stratum of mathematics between the Calculus and infinitesimals and created the new paradox of dividing indivisibles. Although they were more mathematically rigorous, ultimate ratios continued to rely implicitly on the theory of infinitesimals. When criticism of these concepts emerged, Newton's response was that if one agreed with his critics' reasoning, then an ultimate velocity must not exist at all. There would be no final, indivisible velocity that a body reached before it arrived at its final resting place. Indeed, this seemed reasonable enough—that a body would reach its least velocity, or ultimate velocity, before it ceased moving altogether. Unfortunately for Newton, the phenomenon of an ultimate velocity could not be tested, observed, or represented geometrically, keeping him from providing an acceptable proof for defining the notion of the infinitesimal.[47] Newton dealt with this problem in two ways: he strategically made his work difficult, and he used ambiguous language.

Newton struggled with the use of infinitesimals; at the same time that they were the crux of his Calculus, giving him insight into so much of nature, they contradicted his belief in empiricism and modern science. Moreover, it is not clear whether Newton himself knew how to describe the concept in the language of seventeenth-century mathematical science. This, coupled with his loathing of criticism, caused him to rewrite his work in terms of purely mathematical propositions, strategically making the text more difficult to consume and thus narrowing his audience. In the final edition of the *Principia* he wrote,

> Upon this subject I had, indeed, composed the third Book in a popular method, that it might be read by many; but afterwards, considering that such as had not sufficiently entered into the principles could not easily discern the strength of the consequences, nor lay aside the prejudices to which they had been many years accustomed, therefore, to prevent the disputes which might be raised upon such accounts, I chose to reduce the substance of this Book into the form of Propositions (in the mathematical way), which should be read by those only who had first made themselves masters of the principles established in the preceding Books.[48]

By making his texts strictly mathematical, Newton narrowed his audience considerably, allowing only the brightest and the most expert to penetrate his complex propositions. Note too how the conventional rhetorical mobilization of math to obfuscate and narrow his audience is in direct response to the infinitesimal's translative rhetorical force, which exceeded the bounds of Newton's own agency and gave rise to controversies he wanted no part in yet was forced to navigate.

Beyond the strategic difficulty with which Newton crafted his work, he also used ambiguous language. As previously mentioned, infinitesimals were indeterminate concepts. Newton employed indeterminate language because it was philosophically the most fruitful for him, dispensing with the assertion that he was somehow confused, careless, or inconsistent. A quick glance back to his treatment of infinitesimals provides several examples of his ambiguous language. Consider that he defined infinitesimals at first to be those "just nascent principles of finite magnitudes," but what are "just nascent principles?" Should we interpret "nascent" to mean initial or beginning or fundamental, and in each case does not the phrase have different meanings? To argue that this ambiguity was strategic, however, would be to assume that Newton had a greater understanding of infinitesimals than he demonstrated. It is more reasonable to say that Newton used the language at his disposal to explain a novel concept in the best ways he could while preempting criticisms from others.

Newton's discursive shifts and turns regarding the infinitesimal responded to his critics, but they also responded to the infinitesimal's inconsistency with Newton's own foundationalist epistemology. With the use of infinitesimals, Newton was confronted by the limits of his own Platonically informed perspective. The rhetorical force of the infinitesimal effected an undercutting of seventeenth-century foundationalism that Newton and other intellectuals of the time found troublesome. Newton could readily see the pragmatic importance of infinitesimals, but he could not frame them within the boundaries of his own epistemology. Thus, his consistent effort to move away from the explicit use of infinitesimals testifies to Newton's self-censorship and the critiques of his contemporaries. The logic of the infinitesimal was excessive to seventeenth-century scientific discourse, and that excess disturbed Newton perhaps more than any other thinker of the late 1600s.

In any case Newton negotiated the play of infinitesimals throughout his career, taking their use for granted early on and later responding to inconsistencies in his own philosophy and to the criticisms of his contemporaries. Although he did not engage critics readily, this analysis has shown a transformation in language that reveals the impact of external and self-criticisms on Newton and on the ensuing revisions of his use of infinitesimals that later mathematicians inherited. These revisions were rhetorical in nature, meant ultimately to appease or persuade either his critics or his own foundationalism.

Having investigated Newton's rhetoric concerning infinitesimals, it would be appropriate, before discussing the impact of infinitesimals and the Calculus on subsequent mathematical thought, to look at Leibniz's treatment—that is, at how Leibniz dealt with the use of infinitesimals in his Calculus and whether his formulations were in line with, or contradictory to, those of Newton. Leibniz, being more philosophically inclined than Newton, wrote an incredible amount about the use of infinitesimals in the Calculus. He readily engaged the criticisms of his contemporaries through correspondence and published articles. Despite his philosophical interests, Leibniz, much like Newton, displayed an indecisiveness regarding infinitesimals, "considering them variously as inassignables, as qualitative zeros, and as auxiliary variables."[49] He, too, confronted the challenges of empiricism and foundationalism, which demanded observation, experiment, and geometric representation, and he struggled to justify the use of infinitesimals in the language of seventeenth-century science and mathematics.

Leibniz's first article in *Acta eruditorum* explaining his Calculus was, following the model of Descartes's *La géométrie*, bare of proof and justification. Reflecting on the discovery of the Calculus, Leibniz recalled reading Blaise Pascal's *Traité des sinus du quart de cercle* and having a light suddenly burst on him, namely, that one could find the area under a curve by summing up an infinite number of rectangles composing that area. Unfortunately, this same light did not illuminate definitions and proofs for the concept of the infinitesimal that he so obviously employed. Although he struggled to define infinitesimals in his work, "he did not wish to make of the infinitely small a mystery, as had Pascal."[50]

In the early going, then, Leibniz conceived of an infinitesimal "not as a simple and absolute zero, but as a relative zero, . . . that is, as an evanescent quantity which yet retains the character of that which is disappearing."[51] For Leibniz the notion of an infinite convergence toward zero allowed for infinitesimals. To furnish a foundation for this concept he created a general heuristic under the principle of continuity. He noted that "the universality of this principle in geometry soon informed me that it could not fail to apply also to physics, since I see that in order for there to be any regularity and order in Nature, the physical must be constantly in harmony with the geometrical."[52] With this statement, the latter part of which Plato might have written with his own hand, Leibniz conveys not only his faith in the popular mathematical metaphysics of the time, which preached a natural harmony that only geometry could represent, but also his belief in a universal continuity unifying all things.

With the belief in continuity, Leibniz confidently made his position on infinitesimals clear: "I take for granted the following postulate," he stated matter-of-factly, "in any supposed transition, ending in any terminus, it is permissible to institute a general reasoning, in which the final terminus may also be included."[53] Imagine, then, that you have two points on a line, and one point is greater than the other. If the lesser point approaches the greater until both numbers are equal, it is *reasonable* to argue that in the *instant* before the two numbers become equal, a final, ultimate quantity would be transgressed. Essentially, this was the intuitive logic behind Leibniz's use of infinitesimals. Unfortunately for Leibniz, although this reasoning appeared sound, the existence of an infinitesimal quantity was *untestable, unobservable*, and *unrepresentable geometrically*. It was built on "general reasoning" and not on empirical science or Euclidean geometry. Leibniz suffered the same criticism as Newton; he could not prove in modern terms the existence of infinitesimals.

Interestingly enough, his response to criticism was quite the opposite of Newton's. Whereas Newton had started his career using infinitesimals but ultimately chose to disavow them, Leibniz began with the notion of finite differences explained here, only "to be confirmed in his use of infinitesimal conceptions by the operational success ... [of] his differential method."[54] This difference had more to do with contrasts in taste than differences in procedure. For Newton the scientist, infinitesimals indicated a defect. For Leibniz infinitesimals played an integral role in both mathematics and metaphysics. His principle of continuity, which unified all things, required the notion of infinitesimals. Leibniz's adamant stance on infinitesimals caused him to engage the criticisms of his contemporaries and to challenge his own assumptions.

Leibniz responded to the initial attacks on infinitesimals in the mid-1690s by warning against excessive precision that might block intellectual advances and inventions. For him infinitesimals "signify nothing but quantities which one can take as great or as small as one wishes in order to show that an error is less than any which can be assigned—that is, that there is no error."[55] Here we see Leibniz responding to criticisms concerning precision.[56] His critics were requesting some sort of rigorous, precise definition for infinitesimals, but "Leibniz felt that the justification for his Calculus lay in the ordinary mathematical considerations already known and used."[57] What Leibniz gave, then, was a cursory response, as if the existence of infinitesimals was self-evident. When pushed, he had recourse to the reasoning implicit in the use of infinitesimals, arguing that "if one preferred to reject infinitely small quantities, it was possible instead to assume them to be as small as one judges necessary in order that they should be incomparable and that the error produced should be of no consequence, or less than any given magnitude."[58] Leibniz's critics attacked this response as a simple rewording of his nonchalant definition. The problem, they claimed, lay in his use of "incomparable" quantities, which were somehow "less than any given magnitude" yet still retained all the characteristics of finite quantities.[59] It was an obvious contradiction to argue that a quantity was so small so as to be unquantifiable but should still be treated like any other quantity.

In the face of this contradiction Leibniz stood firm, at least initially. When asked to explain the transition from finite to infinitesimal magnitudes, "he resorted to [his] quasi-philosophical principal known as the law of continuity."[60] Leibniz set this principle on high, claiming that it "might help establish a genuine philosophy which rises above the imagination to seek the origin of phenomena in intellectual regions." Indeed, when his confidence in infinitesimals peaked, Leibniz wrote a letter (in 1692) to another aspiring scholar who had communicated reservations about using infinitely small quantities. Leibniz counseled him thus: "Sir, lay aside your fears about the tortoise that the Pyrrhonian skeptics have made to move as fast as Achilles. You are right in saying that all magnitudes may be infinitely subdivided. There is none so small in which we cannot conceive an inexhaustible infinity of subdivisions. But I see no harm in that or any necessity to exhaust them. A space infinitely divisible is traversed in a time also infinitely divisible. I conceive no physical indivisibles short of a miracle, and I believe nature can reduce bodies to the smallness Geometry can consider."[61] Here we see Leibniz at the height of his confidence in infinitesimals. By the end of the seventeenth century, however, he was less emphatic. None of his explanations provided the rigorous definition his critics were seeking. Over time the criticisms mounted, and Leibniz's resolve wore down.

Lacking a majestic definition for infinitesimals, Leibniz began to search for ways to reorganize his differential Calculus so that they were no longer the focus. Like Newton, he argued that infinitesimals were not of consequence. Rather, their *ratios* were what allowed for the Calculus to model continuous motion. Although this reorganization relied implicitly on the existence of infinitesimals, it effectively appeased several of his critics and allowed the elite scientific societies to accept the Calculus without explicitly contradicting their own tenets. The problems seventeenth-century mathematicians had with the Calculus were not with its practical ability to solve difficult problems but with its break from the traditional canon of mathematics handed down by Euclid. Emphasizing ratios rather than infinitesimals effectively removed the latter from immediate purview; it was a kind of mathematical sleight of hand. As José Benardete notes, the infinitesimal "is later expelled from mathematics, by Gauss and Cauchy, not on any ontological grounds, not through any specific concern with nature and the world, but expressly in the name of rigour."[62] In essence this reorganization of mathematical symbols by way of ratios was more acceptable and thus more *persuasive* to the intellectuals of the time.

This is not to suggest that the criticisms Leibniz endured had no other impact. Near the close of his career, Leibniz once admitted that he "did not believe at all that there [were] magnitudes truly infinite or truly infinitesimal" but that these were "fictions useful to abbreviate and to speak universally."[63] By the end, then, Leibniz had acknowledged the rhetorical features of the infinitesimal. The infinitesimal was truly a useful fiction. Although he continued to argue that concepts like the infinitesimal "must not be taken too strictly or literally," his adamant stance on the existence of the continuum and of infinitesimal quantities outside human reason softened.[64] This softening, however, was not the legacy for which he was remembered. Leibniz's legacy was the continuum, continuity, unity, harmony, and the staunch belief that the Calculus, through the use of infinitesimals, could bring insight into God's coherent cosmos.

The Rhetorical Force of the Infinitesimal

By the end of their careers, Newton and Leibniz had "systematized the methods of solution of a tremendous variety of problems, methods based on the use of infinitely small and infinitely large magnitudes."[65] The Calculus was truly amazing; with it one could determine trajectories, movements, velocities—all the motions and orbits of the cosmos were determinable under the powerful Calculus. Even so, the Calculus described motion through an abstract, nonempirical, and non-Euclidean concept. Amid the revolution of modern science and in a sea of foundationalist discourse, this could not help but have a significant impact on subsequent intellectual perspectives.

At first infinitesimals were viewed with awe, as scientists and mathematicians realized their explanatory force. Later, after the initial shock had worn off, intellectuals began to question the integrity of such a concept as the infinitesimal because it did not surrender to trials of empirical or mathematical verification.[66] Newton and Leibniz, as the two central agents of the Calculus, rhetorically negotiated these criticisms in several different and influential ways. Having analyzed their rhetoric, we can now investigate the rhetorical force of the infinitesimal on later scientific and mathematical discourse and, generally, on European thought.

Prior to the Calculus, mathematical discourse had relied primarily on geometric rigor. Equally significant, science in the seventeenth century made a move toward empiricism as modernism took hold of the European intellectual climate. Scientists wishing to establish ethos with their contemporaries were required to conduct rigorous experiments verifying hypotheses drawn from observation. Likewise, mathematicians were expected to prove their theorems according to the rules of Euclidean geometry.[67] The Calculus, however, pushed beyond the limits of "acceptable" scientific and mathematical practice. In using infinitesimals the Calculus did not adhere to the empiricist's criteria for authentication. Nor could one represent infinitesimals geometrically, but the Calculus was so powerful and unique that scientists and mathematicians could not replace or dismiss it. Caught in a paradox, the scientific and mathematical communities were forced to deal with the problem of the infinitesimal's nonrepresentational, nonempirical status. They did so first by requesting a rigorous definition of infinitesimals in the language of modernism. When Newton and Leibniz could not furnish a satisfactory definition, these communities criticized them and their methods. Newton and Leibniz were then forced to negotiate these criticisms at some level.

One strategy used by Newton and Leibniz was to argue that the internal consistency of mathematics trumped the need for empirical verification. They then reorganized their methods to emphasize the *ratio* of infinitesimals rather than the infinitesimals themselves. Here we can detect a significant development in Newton's and Leibniz's discourses. Because infinitesimals were not expressible in geometric form, Newton and Leibniz began to search for other ways to use and explain them. They found that speaking in terms of ratios provided a more rigorous proof for the Calculus. This development would be the beginning of a move in mathematics away from pure Euclidean geometry and toward a more arithmetical form of mathematical proof. Moreover, it would signify a change in mathematical philosophy that persists to this day. As one contemporary mathematician proudly writes, "Mathematics deals with relations rather than with physical existence, its criterion of truth is inner consistency rather than plausibility in the light of sense perception."[68]

The use of infinitesimals did not allow for a geometric representation, forcing mathematicians to seek other methods of proof and driving them away from the spatial relations on which they had relied for so long. However, the use of infinitesimals was not fully accepted in the mathematical community until Augustin-Louis Cauchy's system of limits in the 1820s finally provided a complete arithmetical treatment. This was one of the final steps away from Euclidean geometry as the dominant form of mathematical reasoning, and in part Newton's and Leibniz's rhetorical negotiation of infinitesimals allowed for the departure.

The influence of the infinitesimal went beyond the mathematical community; the scientific community also had a stake in the development of the concept. Perhaps even more than for mathematicians, infinitesimals were a hard pill to swallow for seventeenth-century scientists. In no way could one observe or test the existence of the infinitely small, an obvious contradiction to the tenets of empiricism. Newton's and Leibniz's rhetorical response to empiricists' criticism was to suggest that the senses can misrepresent the truth of things, that clearly logical concepts should not be held back by our own pedantic perspectives, and that ideas can sometimes be more powerful than methods. As Ian Stewart argued, the criticisms of George Berkeley and other seventeenth-century intellectuals "were spot on, but people kept using calculus because it always gave sensible answers. It was more like magic than mathematics, but the spell worked."[69] The "magical" productivity of the infinitesimal forced the empiricists' hand, generating a more open-minded approach to other ways of knowing. The concept of the infinitesimal shook Francis Bacon's dogmatic empiricism.[70] Scientists began to violate the tenets of foundationalism by trusting more fully their own reasoning and creative intuition. In fact, mathematicians later would realize that a system's integrity should not rely so much on what was "discovered out there" as on what *we* "created in here." Via the infinitesimal, scientists and mathematicians recognized the problems with strict foundationalism and made room for something other than observation, allowing nonrepresentational concepts and intuitive argument to play important roles in the production of scientific and mathematical knowledge.[71]

As large a role as the constitutive rhetoric of the infinitesimal (by which I mean the apparatus of interrelated argumentative structures giving presence to the infinitesimal) played in subsequent mathematical and scientific discourse, we should not neglect its influence in other areas. The infinitesimal would prove fundamentally important in several other European intellectual circles. In philosophy, for example, thinkers began to integrate the Calculus into discussions of metaphysics and ontology, allowing "the fantasy of a transcendental origin, an ultimate guarantor of Truth unsituated in time, space or history" to shape the intellectual climate of eighteenth-century Europe.[72] The Calculus proved to be an "indispensable ingredient in the whole complex of eighteenth-century optimism."[73] In particular, Leibniz's principle of continuity had a tremendous impact, allowing for the emergence of much of the eighteenth century's philosophical optimism, belief in cosmic harmony, and drive toward unity.

At the same time, intellectuals began to relate infinitesimals to theology. By the mid-eighteenth century parts of the Bible were being interpreted as supporting "the infinity of the universe and the actual multitude of worlds."[74] Richard Bentley employed Newton's *Principia* to prove the existence of God, to show "that the order of the universe could not have been produced mechanically" and thereby to give believers an assurance in Creation.[75] Bentley's method of proof became a model for future advocates of the divine, and although the church did not receive these interpretations kindly, they illustrate the significant impact the infinitesimal and the Calculus had on European intellectual climates. The infinitesimal became a concept larger than those who conceived it. Ironically, infinitesimals took (and still take) on monumental proportions, simultaneously offering comfort and distress, depending on one's perspective.

Given all these influences, perhaps the most significant mark of the infinitesimal on European thought was on the episteme of representation. By the middle of the 1600s representation dominated European intellectual thought.[76] Empiricism demanded of each symbol a referent, something that could be tested and verified or, if need be, properly corrected. Seventeenth-century mathematics was likewise bent on representation, each symbol in Euclidean geometry having a corresponding geometric object to which it referred. Even in philosophical thought, the meaning of truth was dominated by correspondence and correctness, both of which require representation. The concept of the infinitesimal betrayed all efforts at representation familiar to the seventeenth century. Indeed, it was the incorrectness Newton could not bear. The infinitesimal thus opened the way for later thinkers to transgress the boundaries of modernism and consider problems beyond the limits of representation (such as the "impossibility" of imaginary numbers, a topic we explore in the next chapter).

With the invention and incorporation of the infinitesimal into mathematics, Newton and Leibniz deployed a concept that took their work (and Western thought generally) in directions they could not possibly have anticipated or intended. The infinitesimal marks the limit for measuring and comprehending motion, but it also marks the limit of Newton's and Leibniz's intentionality with regard to the Calculus. Newton, Leibniz, and other seventeenth-century thinkers were responding as much to the translative rhetorical force of the infinitesimal as they were to each other. The discursive formations of empiricism, foundationalism, and the episteme of representation were all (in different ways) at odds with the notion of infinitesimals, rendering the concept an excellent site for studying the constitutive and situational rhetoric at play in mathematical discourse.

Although this discussion has not exhausted the topics in which the concept of infinitesimals played a role, it has established an understanding of the rhetoric constitutive in the emergence of the Calculus *and* the rhetoric of seventeenth-century intellectuals responding to the force of the infinitesimal. Through the analysis of infinitesimals one sees the ways that rhetoric plays a role in the generation, dispersion, and evolution of mathematical discourse. N. Ya Vilenkin named the Calculus "one of the most remarkable creations of the human mind," and now one can see that rhetoric lies at its heart.[77]

Implications

During the seventeenth-century Newton and Leibniz were involved in what Martin Heidegger called a "mathematical project," a movement toward "mathematizing" nature.[78] This move marked a significant shift in the way human communities perceived and operated in their world. A catalyst for this transformation was the use of infinitesimals, an abstraction built on the rhetorical arguments of seventeenth-century reason and intuitive conjecture. The infinitesimal could be called ghostly, evanescent, nascent, or any number of terms descriptive of *that which departs in its arrival*, but it could never be described as tangible or materially present. Rather, the infinitesimal finds its "substance" in the rhetorical arguments circulating around it—the appeals to logical intuition that transgressed the boundaries of mathematical thinking in the seventeenth century and gave way to so many important changes in Western thought. The radical edge of this claim derives not from the notion that a scientific or mathematical concept can transform thought but from the claim that such a concept is wholly rhetorical. The essay by Davis and Hersh begins an inquiry into the rhetoric of mathematics, but they do not attend to the constitutive rhetoric that emerges with the invention of novel mathematical concepts.[79] By focusing on the Calculus and the concept of the infinitesimal, this chapter has examined the constitutive rhetoric concomitant with mathematical invention, showing not only that mathematics has rhetorical features but also that rhetoric can be constitutive of extremely productive mathematical concepts. Far from enemy or adversary, rhetoric in this sense functions as an engine of and for the growth of mathematics.

The implications of such a claim are significant: the nonrepresentational and nonempirical status of the infinitesimal gave rise to a unique set of arguments novel to seventeenth-century mathematical and scientific practice. The arguments *constituting* the concept of the infinitesimal troubled the simple division between what is "supplement" and what is "real" in mathematics. The infinitesimal raised questions regarding the adequacy of Euclidean geometry for describing motion (or rates of change); it ultimately pointed to the limits of a foundationalist perspective for understanding both the universe and our capacities within it. After Newton and Leibniz the infinitesimal disappeared from mathematics, replaced by a more cumbersome mathematics of limits developed by Carl Friedrich Gauss and Augustin-Louis Cauchy in the name of rigor. Only recently has the infinitesimal reemerged in mathematics in the work of Abraham Robinson, who created a new number theory called nonstandard number, which places infinitesimals on a continuum with themselves.[80] These nonstandard numbers (and what is called nonstandard analysis) have been very useful of late in solving myriad scientific problems.[81] The point is this: it took approximately two thousand years for mathematical theory to begin to release its grip on the ideology of foundationalism (one version of Platonic realism), and it was the rhetorical force of the infinitesimal that presented most forcefully to mathematicians and scientists the limits of such a perspective.[82]

To understand the rhetorical dynamism of infinitesimals, this chapter has approached mathematical rhetoric without an emphasis on the substance/appearance dichotomy handed down by our Greek ancestry. This dichotomy makes it more difficult to see rhetoric as constitutive of mathematical discourse because it presumes a two-world binary that positions all symbolic-action as secondary to whatever substance it conveys. Such a perspective shackles language to a logic of representation in which symbols functions as mediators between subject and object. In focusing on an important mathematical concept and investigating its constitutive rhetoric, this chapter attempts to explode such substance/appearance binaries; it shows that the rhetorical force of the infinitesimal operates beyond the threshold of representation, thus troubling any simple substance/appearance relation. The upshot of such analysis is the ability to analyze the constitutive rhetoric of mathematical invention, to view mathematics not as a language "true" to "reality" but as a translative rhetorical force that increasingly enhances our abilities to reconfigure the social-material world. As Richard Rorty points out, "That Newton's vocabulary lets us predict the world more easily than Aristotle's does not mean that the world speaks Newtonian."[83] Mathematics does not seek out the truth lying dormant in nature; it seeks out patterns of relations, which often become means for negotiating, translating, and sometimes transforming the world humans continuously encounter. As we see in the next chapter, those means are rapidly proliferating.

In the last analysis, one might reasonably ask why moving beyond the confines of Euclidean geometry and the limits of representation correlate so strongly with the increased influence of mathematics. The surface response is that the Calculus allowed humans to accurately model and predict the movement of objects in space, thereby enhancing our abilities to both cope with and transform our environments.

A more nuanced response, however, one attuned to the powers of symbolic action, would add that there is a significant difference between Euclidean and non-Euclidean mathematical discourse; that, prior to the Calculus, mathematics was informed by and tethered to the static representational logic of Euclidean geometry—the still-picture stage of mathematics. Mathematicians had as a result a difficult time modeling and therefore being able to accurately predict dynamic motion. The Calculus introduced the nonrepresentational logic of infinitesimals, limits, and continuums and thus initiated the "motion-picture stage" of mathematics. In the aftermath a new world opened up—a world where dynamic motion became increasingly predictable (see Halley's comet); where relations between gravity, mass, force, velocity, and acceleration became increasingly "known" (meaning understood in their dynamic relationality); and where mathematicians and scientists could increasingly *play* with those relations, reconfigure them, and test how hard or soft they might be. Out of such testing and experimenting uncountable innovations emerged, underscoring again that math does not merely discover reality (or the relations that compose reality); it multiplies reality.[84] And that capacity to expand the real will be placed in even sharper relief with the rise of imaginary numbers, which emerged in the Calculus's considerable wake.

4

How Imaginaries Became Real

Consider the previous chapter a beginning: a first step in rethinking the relations between rhetoric and math. In that chapter we see how the core concepts of the Calculus contradicted the foundations of Euclidean geometry, pressing against and eventually beyond the boundaries of the episteme of representation that had shackled mathematics for so long to a still-picture paradigm incapable of modeling dynamic motion. Out of that rupture came many things, not least of which was the rapid evolution of math itself. And while the previous chapter means to show the role of constitutive rhetoric in the invention of new mathematical ideas, and thus do away with the naive notion that math and rhetoric are antithetical, this chapter focuses instead on the evolution of mathematical discourse and, with that evolution, the corresponding increase in math's translative rhetorical force.

There are perhaps countless examples of math's translative rhetorical force, the Calculus not least among them, but no event I know of puts the powers of mathematical discourse on more spectacular display than the emergence of imaginary numbers—in mathematical notation, i, which equals the square root of -1 ($i = \sqrt{-1}$). Newton never approved of their use and called them "impossible"; Leibniz too was dubious, describing them as existing somewhere "between being and not being."[1] Their eventual emergence and acceptance, however, gave birth to the world many now take for granted. Of their importance Simon Singh observes, "Imagine a world without electric circuits. No circuits, so no computers. . . . And while engineers need the imaginary number to analyse electrical waves, physicists need it to calculate the fundamental forces that govern our Universe."[2] How did this all come to pass? Why did the best mathematicians of the eighteenth century reject imaginary numbers as sophistry, and how did they eventually gain legitimacy just a century later? One way to address such questions is to trace the discursive genealogy of imaginaries, to see how they moved from the margins to the core of mathematical practice—not, that is, simply what mathematicians said about them but, more important, what they did with them and *how they behaved* in their uptake and networking with other mathematical systems, a process that enabled whole new realms of the real to emerge. This genealogy, I argue, is of crucial importance, for it reveals exactly how mathematical discourse fabricates the real, transforming an impossibility into a reality and in so doing expanding the network of mathematical relations such that a new world opens up, one where light travels through wires and sound through the air.

At stake in this analysis is a deeper understanding of the relationship between rhetoric and mathematics, one not so much at the level of argument as in the emergence of the vinculum. Derivative of the Latin *vinclum*, the English use of vinculum emerged in the mid-seventeenth century (interestingly enough, around the same time as the Calculus) to signify a bond or tie (e.g., *vinculum matrimonii*—the bond of marriage). The word has since been taken up by several fields: in anatomy, a connecting band of tissue; in mathematics, a sign indicating several mathematical entities be treated as singular. What I endeavor to show in tracing the genealogy of imaginary numbers is that rhetorical force, whether of mathematical discourse or some other symbolic, lies in the vinculum, in the capacities for address that symbolicity affords, be that the invention of a bond or the severing of one, the strengthening of connective tissue or the weakening of said tissue, the weaving of a network or its deliberate unraveling. These powers are put on astonishing display in the formation of i, where one can begin to see the congealing of mathematical agencies (discursive, human, nonhuman) in the fabrication of the phenomenon we now call imaginary number.

Representation and Imagination

What does it mean to say the logics of representation constrain mathematical thought? One of the best ways to address such a sprawling question is via demonstration. Consider, for example, that mathematicians thought the square roots of negative numbers were both impossible and absurd at least until the eighteenth century, with a fair number of mathematicians carrying that perspective deep into the nineteenth century. The story of why begins, interestingly enough, with negative numbers themselves and offers much insight into the strengths and limitations of representation as a governing episteme.

Negative numbers are so taken for granted today that their once controversial nature sounds farcical. Yet for millennia all the best mathematicians "rejected negative numbers as being without meaning because they could see no way physically to interpret a number that is 'less than nothing.'"[3] In Diophantus's *Arithmetica* (widely considered the arithmetic counterpart to Euclid's *Elements*), for example, he considered only positive roots when solving quadratic equations because negative roots made no sense—a perspective that held for approximately two thousand years.[4] Let's pause and consider the phrase "made no sense." In this case, for most mathematicians up through the Renaissance, negative numbers "made no sense" because they were literally nonsensibles; they stood outside the logic of representation that assumed numbers were *magnitudes* whose relations could at least in theory be *demonstrated* via plane geometry. What could it possibly mean, we can imagine an ancient mathematician asking, to have −2 olives? For them it could mean nothing at all, for the meaning of numbers came from their representation of the magnitude of things-in-the-world and the capacity to then place those magnitudes into relation with one another arithmetically (or geometrically, if our example was geometric).[5]

The rejection of negative numbers in this way enjoyed near-universal adherence until the Renaissance. Up through the sixteenth century, the great majority of mathematicians were uncomfortable with negative numbers, seeing their existence as more difficult to prove (both theoretically and pragmatically) and thus potentially threatening to math's claim as the language of the *kosmos* (a claim of many a mathematical realist since at least the Pythagoreans).[6] Descartes's work played a significant role in easing these concerns, as his Cartesian plane simultaneously made negative numbers more "real" (spatially equivalent to positive real numbers) and at the same time incredibly useful for solving mathematical problems. Let's revisit that plane briefly: We already discussed the complex metaphorical content of this mathematical apparatus in chapter 2 and how it functioned as a powerful translation machine between algebra and geometry, enabling the mathematization of countless previously recalcitrant domains. Here, however, I want to examine the symmetrical relation between positive and negative numbers crucial to the plane's translative force. Prior to its invention, numbers were connected almost strictly with the concept of magnitude, which aligned negative numbers with the unreal. Descartes's insight was, in part, to think of numbers spatially as well as in terms of magnitude.

Note how both x- and y-axes in figure 6 create symmetry between the positive and negative numbers such that the negative numbers "mirror" the positives. While a seemingly small detail, this feature of the Cartesian system marks a major

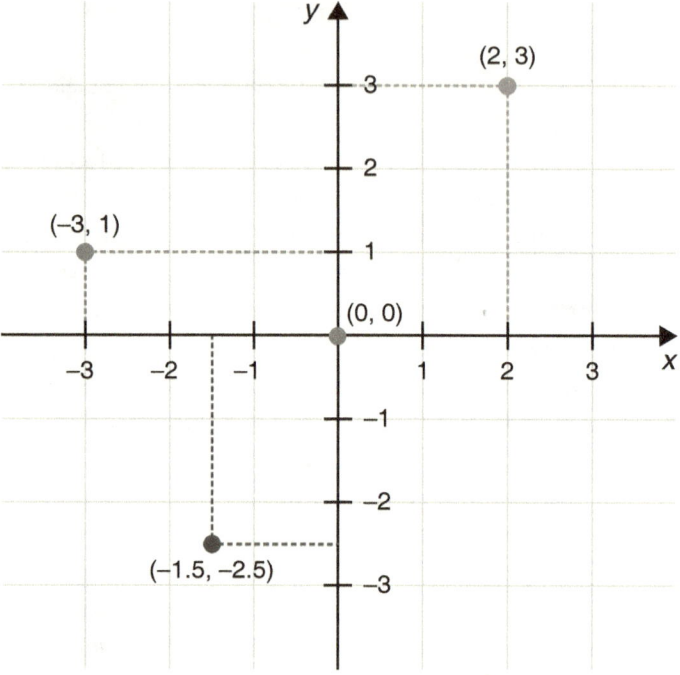

Fig. 6 | Two-dimensional Cartesian coordinate plane

shift in how mathematicians thought of number and of mathematics itself. Prior to Descartes's work, both number and mathematics in general were aligned with Euclidean geometry in part because one could always demonstrate diagrammatically the truth of one's propositions (recall Plato geometrically proving the existence of the eight-unit square in *Meno*).[7] Most mathematicians up through the Renaissance objected to everything from negative numbers to much of algebra precisely because these mathematical "objects" could not be objectively proven—that is, demonstrated diagrammatically according to the laws of Euclidean geometry. Descartes's Cartesian coordinates changed all that: it allowed one to easily translate a complex algebraic statement into a geometric shape, just as one could translate a complex geometric shape into an algebraic statement; it gave spatial presence to negative numbers, treating them as symmetrical counterparts to positive numbers, which allowed for the solving of previously unsolvable problems; and it liberated mathematics to some extent from the diagrammatic paradigm of Euclidean geometry. One quick illustration of this is the fact that today we read "a × a" as a² without even seeing a square of length "a" in our mind's eye. This sense of "making sense" mathematically would, it is safe to say, be completely foreign to Plato and most other mathematicians prior to Descartes.

Despite what the Cartesian coordinate system did to legitimize negative numbers, Descartes still rejected the idea of taking the square root of negatives as useless fictions. To capture this sentiment Descartes gave them the name they carry to this day: imaginary. Why? In large part negative square roots don't behave like real numbers. Recall that part of the metaphorical structure of the Cartesian plane is ordering the numbers spatially along a line, moving out from zero in both directions. This is how real numbers work: they increase or decrease according to the logic of magnitude. When you square a positive whole number greater than 1 it always gets larger, and when you take the square root of that same number it always gets smaller. And for arithmetic to comport with geometry, they must. Thinking geometrically (like the ancient Greeks), squaring a number meant, literally, to construct the square from the original segment of line.[8] A square root, then, must always be less, for it is the length of the side of the drawn square (and here, in the dependence on diagrammatic proof, we can see one of the sources of the episteme of representation in mathematical thought). Square roots of negative numbers, however, don't work this way. Consider the imaginary number $\sqrt{-1}$: if you square it, it becomes -1; if you cube it, it becomes $-1\sqrt{-1}$; if you raise it to the fourth power, it becomes 1. And as you continue, a circular pattern emerges, the value oscillating between 1 and -1. The real numbers are linearly ordered according to magnitude. Imaginary numbers appear instead to be cyclical, at least relative to real numbers. On top of that, using them with other real numbers enables all kinds of absurdities to emerge, scuttling the sense-making power of the reals. For these reasons Descartes and his contemporaries rejected imaginaries as at best nonsensical and at worst a danger to the sense-making power of mathematics.

Once articulated, however, some ideas die hard. And while Descartes was busy writing their epitaph, a few other mathematicians had for some time been using imaginaries, often in secret, to solve the most challenging algebraic problems of the day. The first evidence we have of such use comes from Gerolamo Cardano's *Ars magna* (1545) and regards the solution to the general cubic $x^3 + ax^2 + bx + c = 0$, which can be reduced to $x^3 + px + q = 0$. There are three cases for this cubic, assuming only positive coefficients and values for *x* (this was known as the depressed cubic):

$$x^3 + px = q$$
$$x^3 = px + q$$
$$x^3 + q = px$$

Scipione del Ferro is credited as the first to solve the first case, and perhaps the second and third as well. These solutions were eventually passed down to Cardano, who apparently signed an oath of secrecy, only to later publish them in his *Ars magna* (1545). Why the need for oaths and secrecy? In part this had to do with the competitive nature of mathematics during the Renaissance, where the best mathematicians would often enter competitions in the hopes of winning the goodwill of a wealthy benefactor.[9] As such, novel techniques for solving problems were often fiercely guarded. Equally important, however, was the fact that no mathematician of any renown would admit to using something the mathematical community considered sophistry. And in the second equation there is the possibility of a solution that includes what was perceived at the time as sophistry, namely, the square root of a negative number. Underscoring the discomfort with and perhaps even fear of such cases, Cardano completely avoided discussion of what became known as *casus irreducibilis*, which described all cubic solutions with a negative expression under a radical. Notwithstanding his evasion in this regard, Cardano was the first to introduce complex numbers into algebra, though he had serious concerns (foreshadowing Descartes's) about the mathematical absurdities they could produce (using negative radicals, for example, one could divide 10 into two parts, the product of which is 40).[10]

Despite all the secrecy, fear, and repression, negative radicals did eventually come to light. Rafael Bombelli's *L'algebra* (1572) was the first extant text to openly discuss *casus irreducibilis*.[11] There he described having "what he called 'a wild thought,' for the whole matter 'seemed to rest on sophistry.'"[12] Testifying to the general perspective on negative numbers prior to Descartes, and especially the square root of negative numbers, Bombelli's "wild thought" was simply to imagine how the only real root of $x^3 = 15x + 4$, namely x = 4, might be expressed as $x = 2 + 1\sqrt{-1} + 2 - 1\sqrt{-1}$. This is the first example we have of conjugate imaginary numbers, though their eventual importance could not be seen by Bombelli at the time (for he simply used negative radicals to reexpress a known real solution). Nevertheless, Bombelli's work created more space and legitimacy for negative radicals in mathematical practice. Those radicals, as we know, were later taken up in Descartes's famous *Discourse on the Method of Reasoning Well and Seeking Truth in the Sciences* (1637), where he gave negative radicals their name (imaginary numbers) and associated them with geometric impossibility. Brian Blank summarizes Descartes's view of imaginaries concisely: "For any equation one can imagine as many roots [as its degree would suggest], but in many cases no quantity exists which corresponds to what one imagines."[13]

While other mathematicians engaged with imaginary numbers, including Newton and Leibniz, they remained on the margins for nearly two centuries, where they were put by Descartes and others. In 1693, for instance, Leibniz shocked the mathematical community by transforming a positive real number into an imaginary decomposition: $\sqrt{6} = \sqrt{(1 + \sqrt{-3})} + \sqrt{(1 - \sqrt{-3})}$. Even so, Leibniz's characterization of complex numbers as "a sort of amphibian, halfway between existence and nonexistence" further illustrates the liminal space afforded imaginary numbers at the end of the seventeenth century.[14]

Not until Leonhard Euler and Carl Friedrich Gauss did i move from the margins of mathematics to the core. Recall that, as Steven Strogatz observes, "until the 1700s or so, mathematicians believed that square roots of negative numbers simply couldn't exist."[15] How did Euler and Gauss effect such a profound about-face? Interestingly enough, this part of the story is rather rhetorical—that is, to do with discursive transformation. In what might appear a minor notational invention, Euler introduced the figure i and the equality $i = \sqrt{-1}$. That simple equality had significant rhetorical force, transforming through equivocation the, at best, liminal reality of negative radicals into a *potentially real* mathematical concept. That potentiality was then quickly realized in one of the most famous and celebrated equalities in the history of mathematics: $e^{\pi i} + 1 = 0$. Beyond the significance of this equation for the history of mathematics—as Carl Boyer and Uta Merzbach note, Euler's equation "contains the five most significant numbers (as well as the most important relation and the most important operation) in all of mathematics"—consider what it does for i.[16] Euler's identity effectively positions i in relation with the most important numbers in mathematics at the time: 0, 1, π, and e. More important, however, Euler's equality renders i essential to the mathematical insight the equation captures. What insight is that? At the risk of oversimplifying, this equation is the first to unite a fabulously complex set of mathematical ideas spanning algebra, geometry, logarithms, and trigonometry into one beautifully parsimonious equality.[17] And Euler was able to build this surprisingly simple yet complex mathematical network only with the connective

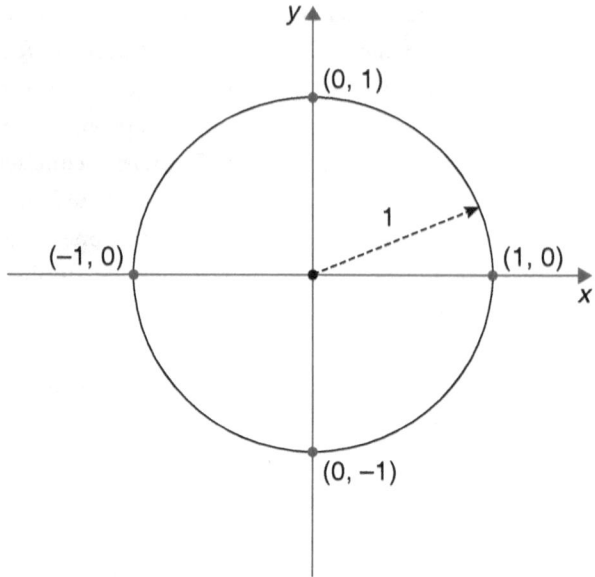

Fig. 7 | The unit circle

tissue (the vincula) *i* provided. What was that connective tissue? Recall the cyclical behavior of imaginary numbers as we increase them exponentially. Now consider the unit circle (fig. 7), from which the concept π comes.

In the unit circle, and in every circle, π is the relationship between the circumference and the diameter—it is the number one always gets when dividing the circumference of a circle by its diameter. Thus π is a number (a constant magnitude) and more than a number (a conceptual relation unique to circles). Now take the unit circle, place it on the Cartesian coordinate system with its center at the origin (0, 0) and slice it in half along the *x* axis. Imagine sliding the top half right along the *x* axis one unit and the bottom half right three units so that the left side of the lower arc lines up with the right side of the upper arc (see fig. 8). The result would be an approximation of the sine-wave function, which may be familiar to those who took trigonometry as part of their secondary education.[18] We need not delve deeply into trigonometry, however, to begin to see the connective tissue forming here—surprising commensurabilities between the circularity of imaginary numbers, π as an expression of the unit circle, and

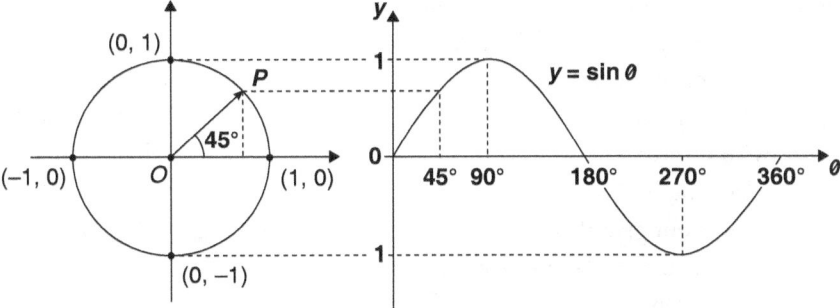

Fig. 8 | Translation of unit circle into sine function

the classic recurrent sine-wave function of trigonometry, which captures visually the recurrence of the numerical value of angles.

Already, even before fully understanding Euler's equation, we can begin to see relations forming that "concept-stretch" the elements involved.[19] Prior to trigonometry, for example, π was strictly defined as the relation between the circumference and diameter of a circle. Descartes's translation machine, however, enabled not only translations between algebra and geometry but also between geometry and arithmetic. In fact, prior to *La géométrie*, angles in mathematics rarely had numerical signifiers: they were not arithmetized. But the numerical design of Cartesian coordinates invited the arithmetic translation of geometry. The result was the rise of analytic geometry and modern algebraic trigonometry, which uses the unit circle to arithmetize the study of angles. Recall the original meaning of π as the relation between circumference and diameter. For the unit circle this means that $C = \pi d$ becomes $C = 2\pi$. Stretching that simple relation to the study of angles, we can see how one might arithmetically reimagine the concept "circumference" as "360 degrees," which would then allow $C = 360° = 2\pi$, which would mean $180° = \pi$, and so on, until you have the periodic sine function shown in figure 8. Note, however, that in this context π no longer means the relation between circumference and diameter. Instead, it means the movement of a point (like point "P" in fig. 8) along the circumference of a circle (e.g., radially), that point traveling the distance π as the angle of degree increases from 0 to 180. As Lakoff and Núñez observe, "Where π was previously only the ratio of the circumference of a circle to its diameter, 2π now becomes a measure of periodicity for recurrent phenomena, with π as the measure of half a period. This is a new *idea*: a new *meaning* for π."[20] And this is one of the primary ways mathematics evolves—through concept-stretching a previously established relation into a novel hybrid, the concept involved becomes richer, more polysemous and rhetorically dense, with a broader network of connections that open the network of relations we call mathematical reality to difference.

Already in our genealogy of *i* one can see significant mathematical evolution. From negative numbers *becoming real* through the translative force of the Cartesian system to the concept-stretching of π in the arithmetization of the unit circle, which gave rise to a whole new area of mathematical study.[21] Just on the horizon we can also see the emergence of a whole new way of thinking mathematically: not linearly or geometrically but around and through the concept of recurrence, which is, not by chance, closely linked with the concept of current (e.g., electrical current, ocean current, etc.). But we get ahead of ourselves. We must first fully understand the discursive realization of *i* before turning to its many impacts on material culture.

Returning to Euler's famous equation, $e^{\pi i} + 1 = 0$, one might now be able to see the critical linkage between π as a recurrent period in trigonometry and the recurrent movement of *i* in exponential form. But what of *e*? Put as simply as I'm able, *e* is a number that comes from the study of logarithms, which is the mathematical practice of mapping one mathematical system (like addition) onto another (like multiplication). These logarithmic transformations can also be used to study rates of change, which was the central focus of Newton's and Leibniz's Calculus. It turns out that if one wants to find a function that maps addition to multiplication in such a way that the rate of change is identical to itself, the number that emerges is *e*, or 2.718281828459....[22] What does it mean, then, to raise *e* to the exponent πi? Recall that through Cartesian coordinates mathematicians understand negative numbers spatially as the inverse of the positives. One can also think of this rotationally: to multiply by a negative number, in other words, is to *rotate* on the Cartesian coordinates 180° (this is why when we multiply a negative by a negative we get a positive). We also know that logarithmic functions map between systems, in this case between addition and multiplication, and that we should be thinking, along with Euler, of π in terms of its trigonometric meaning. Gathering all those thoughts together allows for the translation: $e^{\pi i} = \cos \pi + i \sin \pi$. Note that we said translation and not *equality* here. That's because the sign "=" here means more than equivalence. It means a logarithmic mapping from one system (exponential logarithms) of mathematics to another (trigonometry). And if one simplifies the right side of the equation trigonometrically, one would observe that cos π = −1 and sin π = 0, which means that i sin π = 0, which means $e^{\pi i}$ = −1, or $e^{\pi i}$ + 1 = 0.

Recurrence and Translation

We now see how Euler got to his famous equality, and we can see at least some of the vast mathematical network that equality implies, but why does any of this matter beyond the world of esoteric pure mathematics? It matters, in part, because in the details, however gray and meticulous they might appear, we can trace the discursive construction of "the real." Prior to Euler's work, imaginary numbers were considered impossible, and *they were impossible* within the constrained world of Euclidean geometric representation. Only later, through the work of Gerolamo Cardano, Rafael Bombelli, John Wallis, Leonhard Euler and many others, were imaginaries made real, not by their a priori objective status but by virtue of the network of connections and translations they enabled between geometry, arithmetic, algebra, calculus, trigonometry, and logarithms. And their slow acceptance as "real" marked a major departure in mathematics from the logics of representation that previously constrained mathematical thought. To put it bluntly, without i there is no Eulerian equality, which means no network of connections between the different fields of mathematics involved, more friction when one wants to translate and concept-stretch mathematically, and more deceleration of mathematical and scientific innovation. The role of i in Euler's work had a tremendous impact on both pure and applied mathematics. Euler's identities involving i, as we now know, could easily be combined with trigonometry to show that an imaginary number, taken to an imaginary power, can be a real number; hence $(e^{\pi i/2})^i = e^{\pi i \wedge 2/2} = e^{-\pi/2}$. And Euler later showed that "there are, in fact, infinitely many real values for i^i."[23]

Now might be an opportune moment to consider what is meant when one says "real." Are imaginary numbers real? Is Euler's i real? If by "real" one means representative of an a priori that is always already there, waiting to be discovered, the discussion quickly becomes a matter of faith. But if one means by "real" a set of relations that compose reality, then we can begin to see how the discursive invention of i created space for a new set of relations to emerge, a set of relations that led to one of the most important equations in the history of math. And here the value of a rhetorical approach to math is placed in sharper relief: namely, one sees more clearly that mathematical discourse has a profound capacity *to make things real*; that the "making real" of a mathematical concept is at least in part a process of rhetorical-discursive invention, a process that can potentially introduce novel relations within mathematics; that those novel relations are the material from which new math is invented; and that that new math, like Euler's equation, introduces new networks of relationality that have the potential to transform both mathematics and the world ontologically—that is, at the level of material existence.

To gain a better understanding of i's translative force, however, we need one more mathematical innovation, which came in the form of a blend, ironically enough, of Descartes's coordinate plane and imaginary numbers. We should also note that while Euler's work did much to legitimize imaginary numbers, there is little evidence that even he fully understood the new reality that i would make possible (much like the concept of the infinitesimal, i has its own rhetorical force). In fact, most mathematicians of his time viewed the Eulerian identity with as much awe as they did understanding, and so imaginary numbers remained on the margins until Gauss offered a more complete treatment of complex numbers (reals combined with imaginaries in the form $a + bi$), which he used to complete the mathematical system of number theory (as the fundamental theorem of algebra proves).[24]

Gauss's results, along with Euler's, would eventually secure a place for imaginary numbers within the mathematical community.[25] And as imaginary numbers became more widely accepted, mathematicians felt freer to openly experiment with them. That freedom led to the visual-conceptual blend called Argand diagrams (better known today as the complex plane), the ideas for which were independently developed by Caspar Wessel and Jean-Robert Argand around the same time Gauss's *Disquisitiones arithmeticae* (1801) was published (Wessel in 1797 and Argand in 1806).[26] Knowing what we now know about imaginary numbers, let's look more closely at the Argand diagram (see fig. 9). The first thing to note is the blending of the Cartesian plane with imaginary numbers, but what does that mean? In replacing the y axis with an imaginary axis, what is happening at the symbolic-discursive level? What ideas are lurking in this diagram and how might it shape the thought of those that *see through it*? The first and most obvious point, building on our commentary, is that "taking this simple step," as Boyer and Merzbach note, "made mathematicians feel much more comfortable about imaginary numbers, for these now could be visualized in the sense that every point in the plane corresponds to a complex number, and vice versa. Seeing is

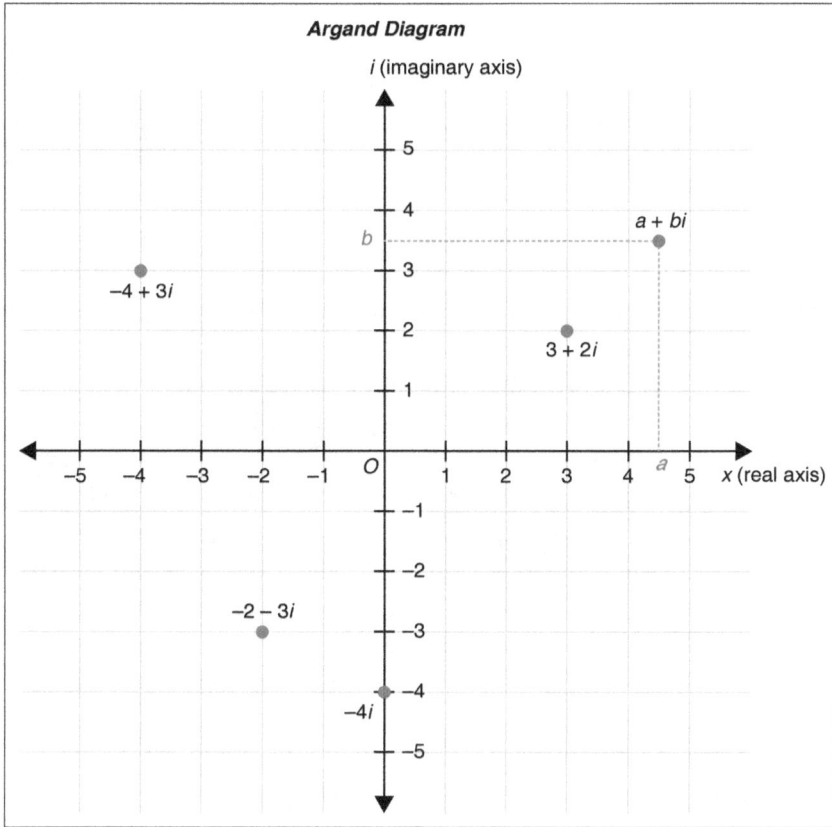

Fig. 9 | Argand diagram (complex plane)

believing, and the old ideas about the nonexistence of imaginary numbers were generally abandoned."[27] Although Boyer and Merzbach oversell how quickly mathematicians abandoned their "old ideas" about imaginaries, their point about visualization is astute, not simply because mathematicians could suddenly "see" complex numbers, but, more important, because the Argand diagram further extended the network of relations of *i* in mathematical practice. The Argand diagram concept-stretched the Eulerian *i* into yet another novel mathematical-visual context, giving birth to a new hybrid in the process.

That new mathematical hybrid asks its user to see the world through new eyes, but to fully understand the new perspective and its consequences we must return briefly to another feature of the Cartesian plane. Recall that a basic conceptual element of Cartesian coordinates is placing numbers on a "line" spreading out symmetrically from zero. That symmetry proves crucial, not just for the legitimization of negative numbers but for the extension of imaginaries as well. Consider: on the Cartesian plane not only can one plot lines and curves and translate between geometry and algebra, but one can also rotationally transform positive points or objects into their negative counterparts. To do this, as we already observed, one simply multiplies by −1 to affect a 180-degree rotation. Yet what if we want to rotate an object only 90 degrees? The kernel of insight—the idea at the heart of the Argand diagram—is that, instead of thinking only spatially with the Cartesian plane, one can also think *rotationally*, and, once one is thinking rotationally, one begins to see the potential of blending imaginary numbers with real numbers to create a rotational vector plane. For instance, on the Argand diagram plot x = 3. Now multiply by i, and one can easily see how that operation moves the original point 90 degrees (or transposes it from (3, 0) to (0, 3i)).[28] Equally important, if one wants to rotate an object 180 degrees, one multiplies by i twice, or i^2, which means the Argand diagram's rotational properties are *commensurable* with the Cartesian plane yet enable additional subtlety.

The real subtlety of the Argand diagram, however, lies in how it encourages one to think both spatially and rotationally at the same time. This is why complex numbers turn out to be so useful for studying the multidirectional rotational motion of wave-like phenomena, whether ocean waves or water flowing around an object or electromagnetic waves. Each of these phenomena simultaneously "expand" (move away from an origin) and rotate, both parts of which complex numbers can efficiently mathematize (the real for expansion, the imaginary for rotation). It would be difficult, in fact, to overstate the translative force of imaginary numbers, complex numbers, and Argand diagrams on the contemporary world. Together they had a cascading effect, simultaneously completing number theory; creating whole new realms of mathematical inquiry (vector analysis, linear algebra, and many other forms of math beyond Calculus); providing the mathematical substrate that enabled the electrical revolution of the nineteenth century; and even today forming the mathematical bedrock of our modern communications system. And that is but the tip of a very large iceberg.[29]

Implications

How did imaginaries and their hybrids (complex numbers, the complex plane) open up these new domains, multiplying the fields of power on which humans could play, that play eventually expanding the real irreversibly? There are seemingly infinite results in mathematics, physics, optics, genetics, chemistry, engineering, and so on that imaginary numbers helped make possible, a full accounting of which would take thousands of pages. No one would buy that book, and even fewer would read it!

We need not detail every innovative application, however, to observe the radical shift in discursive practice imaginaries entailed. Consider the Cartesian coordinates discussed earlier: note that, despite the apparatus's great powers of translation, it enables only a one-dimensional static study of mathematical problems. It remains squarely within the episteme of representation, where independent entities like sides, angles, and bases relate to one another, and may even constitute a larger whole such as a triangle, but remain independently distinct. This reflected Descartes's mechanistic representationalist perspective, where words and things were distinct entities and where transcendental being emerged through the capacity for thought (*cogito, ergo sum*).[30] When Karen Barad wrote that "representationalism separates the world into the ontologically disjoint domains of words and things, leaving itself with the dilemma of their linkage such that knowledge is possible," she might as well have been describing Descartes's philosophy.[31] But if one wants to address a nonstatic, multidimensional problem mathematically, one must go beyond both representationalism and Cartesian coordinates, which not by chance is exactly what both the Calculus and imaginary numbers do. They push mathematical thought beyond the boundaries of hypostatic representationalism and into the realms of dynamic multidimensional phenomena. Imaginaries are particularly powerful in this regard because they enable, when combined with calculus, the study of complex entanglements between motion ("rates of change") and recurrence ("rotation"). And it is precisely this entanglement of motion and recurrence that one finds in countless "natural" phenomena, from electromagnetic waves to ocean waves to DNA.[32]

And while it is surely the case that these entities of matter existed prior to James Maxwell's equations or James Watson's and Francis Crick's double helix, that does not mean those *matters* are inert, just waiting to be discovered by some special genius. No, they are not "objects" in the Cartesian sense but forms of relational intra-action that have congealed into a temporarily stable, but always becoming, matter. Matter in this sense has its own agential force and can be articulated, extended, and transformed through discursive practices. In fact, the very possibility of socializing a new relation into the social collective rides on the structure of discursive practices with which and through which one thinks and scribbles. In this sense *i* is a sign for a broader discursive practice that enables (is a condition for the possibility of) the mathematical articulation of recurrence, the expression of which further entangles humans and nonhumans into increasingly complex and consequential relations. This in part is how discursive practices "produce" material bodies: not by alchemy but through uptake, translation, and manipulation (through acceleration, concept-stretching, fabricated relations, etc.), such that the material-body "light bulb" emerges as a novel hybrid of our understanding of electrical current, something only possible in a world where imaginary numbers are accepted as real.[33]

There is so much about our contemporary world that would not be a reality sans imaginary numbers that it is, ironically, daunting to imagine; i is the consummate principle of composition, giving rise to complex numbers, complex planes, and complex functions. Without those there would be no Reimann mapping theorem, which allows us to understand all sorts of problems about heat flow, electromagnetism, and electricity. Without complex functions, there would be no Fourier transform, which serves as the basis for just about every technology we take for granted: radios, televisions, digital cameras (which use the Fourier transform to compress massive amounts of information onto chips smaller than a thumbnail), CAT scans, and MR scans—technologies used to discover the molecular and atomic structure of crystals, from which humans began to understand the molecular structure of viruses, chlorophyll, insulin, and eventually DNA; and just about every sophisticated modeling technique used to process and sample huge amounts of data, from climate change models to economic forecasting. And if that's not enough, we could not even begin to understand the quantum world without imaginary numbers. "Quantum physics is all about 'you're not quite sure where a particle is until you observe it,'" notes mathematician Marcus du Sautoy, from Oxford University. "That's because actually the particle is existing in this imaginary world, it is described by imaginary numbers, and when you observe it, it sort-of has to collapse onto the measuring line, into the real world, the one-dimensional world."[34]

Like I said, the consummate principle of composition, the meaning of which is at least threefold. In one sense i enables the *composition* of a complex mathematical network linking geometry, algebra, trigonometry, complex numbers, complex functions, and so on; it is the material sign of a newly emergent discursivity of recurrence. In another sense that growing mathematical network enables an understanding of the *compositions* (the relations) that constitute phenomena like electricity or molecules. And in a third sense understanding electrical and molecular composition leads to an understanding of their potential *decomposition*—that is, both how they are composed and how one might decompose them to recombine those more elemental phenomena into novel hybrids (which is exactly what mathematicians and scientists have done for the past two centuries). And this, ultimately, is how discursive practices can mark bodies, become embodied, and materially manifest, and in those manifestations accelerate reality's becoming.

What does the genealogy of imaginary numbers ultimately reveal then? In part at least it reveals the power of discourse and symbol to transform an impossibility into a reality, not through some sudden discovery but through the slow building of a network of mathematical relations that eventually made imaginaries real (connected, interwoven, resistant to challenge, expansive of the real); those imaginaries then became more real with subsequent applications, each of which effectively extended their network of relations. The Argand diagram is one of our exemplars because it significantly expanded reality in the 1800s or, put differently, the set of relations that composed what was called reality in the 1800s. It did so in part by rendering commensurable what had previously been considered distinct phenomena (real numbers and imaginary numbers, ocean waves and electromagnetic waves, etc.), expanding the realms of calculability and transforming our perception, apprehension, and capacities for the manipulation of our environments in the process. Imaginary numbers, complex numbers, Argand diagrams—these are so much more interesting than a priori truths; they are instead principles of composition that already have expanded and will continue to expand our collective social-material realities in radical ways (consider, for instance, our evolving understanding of quantum phenomena).

The genealogy of imaginary numbers tells a beautiful if slightly more humble story of mathematics. Instead of a mystically transcendent language of absolute truth, what emerges is a carefully articulated, dynamically evolving network of relations that begins in this case in a moment of rhetorical invention ($i = \sqrt{-1}$). That notational invention (Euler remains perhaps the greatest inventor of mathematical notation) functioned as a thought-experiment and conjecture for Euler, who was no doubt shocked himself by what such a minor shift in discursive practice ultimately enabled. Through such moments of rhetorical invention the powers of rhetoric, which lie in the vinculum, are unleashed, and previously stable concepts (like number, π, Cartesian coordinates, etc.) can suddenly be concept-stretched into new domains, which simultaneously increases the *rhetorical density* of these core mathematical concepts and opens mathematical reality (e.g., the network of relations that constitute mathematical reality) to difference.

Thus the genealogical tracings of this chapter ultimately amplify the crucial importance of rhetorical-discursive invention in the evolution of mathematics, the ways those practices of invention fabricate symbolic apparatuses useful for thickening human relations with the material world and enabling humans to more powerfully manipulate those relations precisely because these symbolic apparatuses were always already in and of the material world—always already informed by the embodied material experiences of motion and recurrence of the organism we call human (which is itself, as Donna Haraway and so many others have shown, an evolving discursive-material entity).[35] And we will see in the next chapter just how profoundly the "matter of being human" has evolved with the rise of an algorithmic culture impossible to conceive of absent the incredible translative force of i.

5

Algorithmic Culture and Economies of Translation

We have seen in the rise of the Calculus and the fabrication of imaginary numbers a strong correlation between the departure from the logics of representation and an increase in the translative force of mathematical discourse. That translative force is often realized in the twenty-first century through algorithmic implementation. In this chapter we leave the confines of technical mathematics and venture into the borderlands between technical and public spheres, spaces from which the social-material worlds of this century are being increasingly fashioned. The problem is that fewer and fewer citizens, including those that build these algorithms, understand their translative rhetorical force—or, put differently, their capacity to transform social-material relations on a scale heretofore unseen.

Indeed, algorithms have never been more influential, yet our collective understanding of how they transform massive networks of cultural power has not kept pace.[1] This is due in part to anthropocentric narratives that consistently stress human agency and in part to corporate interests who benefit from keeping algorithms proprietary. Whatever the causes, algorithms today operate as black boxes largely inaccessible to the majority of citizens whose worlds they continuously reshape. We do not understand the unparalleled productivity of these algorithms and how they work, nor what they *mean* for the social collectives they transform.[2] We cannot see the parameters of judgment—both human and nonhuman—that constitute them; thus those parameters appear as horizons, always present yet distant and untouchable. Unable to see or understand their judgment parameters, we have no comprehension of what these algorithms assume to be true, nor what might happen if those assumptions are violated. As Ed Finn observes, "We imagine ... algorithms as elegant, simple, and efficient, but they are sprawling assemblages involving many forms of human labor, material resources, and ideological choices. . . . To truly grapple with the age of the algorithm and our growing entanglement with computational cultural processes, we need to take action as scholars, teachers, and most of all performers of humanistic inquiry."[3] This chapter seeks to make the horizons of judgment within algorithms accessible—not only to challenge the positivism and mathematical realism that naively apotheosizes algorithms and algorithmic culture but, more important, to position rhetorical scholars as *critical informants*, intellectuals who can answer Finn's call and open up these black boxes for fellow citizens, examine the hidden assumptions therein, and study how they *actively transform* our social-material worlds.[4]

To begin such an ambitious endeavor (one far beyond the powers of a single chapter), I examine the role of economic algorithms and the mathematical discourse from which they sprang in the 2008 financial crisis. I look specifically at the rise of a particular algorithm known as the Li Gaussian copula, which played a significant role in accelerating the spread of subprime mortgages. Unraveling this copula reveals the constitutive rhetorical force of mathematical discourse—its capacity to invent, accelerate, and concentrate networks—but also the fact that inside every algorithm lies a horizon of judgment that, when used, effectively displaces human agency and practical judgment with the agencies and judgments already baked into the mathematical algorithms employed. Those horizons of judgment, however, are often concealed (both intentionally and unintentionally) within forms of mathematical realism that transform mathematical *models* into "hard technologies" for economic production and extraction. I trace this process within David X. Li's original article, showing how rhetorical analysis can reassemble the horizons of judgment hidden within technical algorithms while simultaneously addressing the material consequences of these algorithms both as forces of displacement and as principles of composition.[5]

The Li copula is our algorithmic exemplar. Close examination of David Li's "On Default Correlation" reveals how analogy can expand the boundaries of mathematical computation within technical economic discourse but also how horizons of judgment within algorithms are simultaneously assembled and concealed. The widespread view that mathematically rigorous algorithms are free (or have been purged) of judgment, in fact, is precisely what allows them to become "technologies" (objects, tools, things) in the economic imaginary, sanctioning the marginalization of human agency and judgment. This chapter traces the material consequences of these processes, ultimately arguing that a rhetorical approach to the Li copula—and algorithms in general—reveals the materiality of mathematical discourse, the ways algorithms function as principles of composition, and how they can displace and marginalize *phronesis* (practical judgment) with forms of technological rationality that, in the case of the Li copula, concentrated the networks of structured finance around a single decision apparatus (the copula), rendering those networks both larger and, contra conventional wisdom, more fragile.[6]

Developing these arguments calls first for construction of a rhetorical approach designed to read complex algorithms. Only then will we be in position to reassemble the horizon of judgment within the Li copula, see the artistry therein, and examine the vast financial networks that emerged and eventually collapsed as a result of the copula's spread. Prior to that, however, I briefly trace research in critical algorithm studies, connecting this chapter with contemporary scholarship and revealing how it aligns with and extends current conversations.

Algorithms, Rhetoric, and Math

Algorithms are rapidly transforming the human experience. Consider, for instance, how practices of remembrance have changed. Prior to the twenty-first century, remembering and forgetting happened individually and collectively through rituals, oratory, photography, monuments, museums, and myriad other forms. Yet with each form of mnemonic practice humans played a central role. In recent years, however, a new mnemonic regime has emerged—a memory practice governed primarily by algorithms, one Neal Thomas describes as a "mnemotechnology" that "embeds a neoliberal logic into memory."[7] The edges of this regime come into focus when an algorithmically selected photograph emerges unbidden on Facebook (or any other social media platform). Why that image? How was it selected for recirculation? While answers to these questions are strategically proprietary, one need not see Facebook's algorithmic code to note how the motives for remembrance have shifted from the social (about ethos, identity, association, dissociation, etc.) to the economic (about how recollection can create surplus value). Likewise, one might note the change in one's own role relative to this algorithmic memory work—from catalyst and inventional resource to recipient and node in a techno-economic network of circulation. As Bernard Stiegler states concisely, "Something absolutely new happens when the conditions of memorization . . . becomes concentrated in a technico-industrial machine whose finality is the production of surplus value. . . . There has today occurred a veritable inversion in the relation between life and media: the media now relates life each day with such force that this 'relation' seems not only to anticipate but ineluctably to precede, that is, to determine, life itself."[8]

The economic motives giving rise to mnemotechnologies seem fairly innocuous until one realizes that learning algorithms curate both sides of the information exchange: both content *and consumer*. Several years ago companies like Google and Facebook began using simple learning algorithms (literally a few lines of code) designed to optimize "click-throughs" on internet content. The concept, according to Stuart Russell, UC Berkeley professor and pioneer of artificial intelligence, was that the algorithms would improve curation and targeting of content for individual users and consumers, increasing clicks on said content and thus increasing the success of and price paid for said content.[9] The engineers did not anticipate, however, that these learning algorithms would begin optimizing users as well as content. The reason was simple: the algorithms learned that users with more extreme political and ideological views were more predictable in their click-through behavior. They thus began to cultivate more extreme views among the user population. As a 2017 article in *Scientific American* notes, the trend in algorithmic artificial intelligence is clearly moving "from programming computers to programming people."[10] These algorithms, both Stuart Russell and Rob Reid observe, are part of the algorithmic architecture undermining Western democracies across the globe. In an interview with Reid, Russell stated, "Effectively a few lines of code is in the process of destroying the EU and NATO and possibly western democracies and enabling the resurgence of fascism."[11] Polarization, it turns out, is more profitable than consensus.[12]

Examples like these are proliferating, underscoring the increasing ontological force of mathematical discourse. That in itself may not necessarily be a problem—the problem lies instead in the asymmetries of power and control that emerge from these practices, producing a culture one might describe as "transparently opaque." By that seemingly paradoxical union, I mean to underscore how algorithmic culture simultaneously enables powerful forms of publicity and surveillance (most social media platforms benefit from both) *and* increasingly numerous obstacles for understanding how decisions happen within algorithmically driven domains. These twin phenomena are of central interest to a growing number of scholars within critical algorithm studies, giving rise to—in Frank Pasquale's memorable phrase—not simply a collection of "walled gardens" but a culture that "closely resembles a one-way mirror."[13] On one side of that mirror stand private corporations and government institutions using sophisticated algorithms to "know" its customers and the citizenry. From this privileged position a population becomes increasingly transparent (tracked, quantified, computed, and controlled). On the other side, however, people encounter numerous obstacles to understanding the forces behind complex decision-making. Those obstacles come in many forms: the technical complexity of mathematical code, established legal precedence that enables proprietary claims, and the recursive evolution of "learning algorithms" that are difficult for even the best mathematicians to decode.[14]

This one-way mirror is rapidly destabilizing traditional social institutions. Representational democracies are under algorithmic attack from abroad. Algorithms producing "fake news" regularly undermine the credibility of a "free press" as a check on demagogic power. Algorithmically governed cryptocurrencies are competing with and transcending Federal Reserve Banks. Centralized educational systems are scrambling to adapt to digital pedagogy platforms. Even our juridical institutions are under extreme pressure from the rapid proliferation of algorithmically enabled forms of surveillance, drone warfare, gene editing, and cloning (to name a few).[15] For many scholars the problem here lies not with the algorithms themselves—those can be beneficial or detrimental—but with the associated acquiescence to an increasingly influential yet opaque bureaucracy. As Cathy O'Neil notes, "We have a total disconnect between the people building the algorithms and the people who are actually affected by them."[16] That disconnect is undermining basic democratic principles, where informed decision-making depends on information access and where privacy grants citizens the autonomy to thwart authoritarianism. "When every move we make is subject to inspection by entities whose procedures and personnel are exempt from even remotely similar treatment," Pasquale concludes, "the promise of democracy and free markets rings hollow."[17]

In response to these challenges, scholars have called for an ethics of algorithms. Efforts are currently underway to create legal standards for privacy and data use as well as greater transparency regarding profiling practices and information filtering.[18] Again, for most scholars algorithms are not the problem. The problem is with their implementation and specifically how that implementation perpetuates subjective biases or exacerbates inequalities.[19] Scholars have thus called for both a priori ethical standards and the creation of an independent auditing agency to address issues as they emerge, something especially important within the context of adaptive learning algorithms that evolve as they process data.[20]

The increasing presence of adaptive learning algorithms, however, places pressure on scholars to go beyond problems of implementation. There are countless studies of how algorithms promote bias, inequality, subjective values, a culture of secrecy, and ideological polarization. Yet nowhere near the same level of scrutiny exists for the *construction* of algorithms. To some extent this asymmetry makes sense: humanists and social scientists are naturally drawn to the sociocultural consequences of algorithms, and at the same time few have the training to unpack technical mathematical code. As a result, the critical literature on algorithms is rich with analyses of implementation problems (promotion of bias, fragmentation of social institutions, spread of positivistic culture) yet thin on study of how algorithms are built. This observation is hardly new: as Marc Lenglet notes, "When we want access to what they precisely make ... we face a black box that we usually fail to understand thoroughly and open completely."[21] That failure has a high cost, since most recommendations for an ethics of algorithms are emerging sans a deep understanding of how complex algorithms are assembled, much less how they evolve.[22] What we need to address this asymmetry is a critical mathematical approach to algorithms, something that can attend closely to both their discursive fabrication and their social-material consequence. And that in part is exactly what this book is meant to provide—a means to analytically map the linkages between algorithmic construction, an algorithm's material manifestation in culture, and the consequences—what I think of half-satirically as the aftermath—of implementation patterns.

As we know from the previous chapters, there are numerous ways to approach mathematics rhetorically (see chapter 2). We need them all in this chapter. Modern algorithms are complex assemblages that demand all our tools of analysis, including a neo-Aristotelian emphasis on the strategic use of math to persuade or manipulate an audience, the spotlight constitutive rhetoric places on the agential force of mathematical discourse, and translative rhetoric's emphasis on math as a practice of discursive-material assemblage.

With its emphasis on strategies of argument and persuasion, the neo-Aristotelian approach renders one highly suspicious of implicit or explicit claims to objective truth. Such habits of mind position one well to question discursive processes that claim to move from conjecture and thought-experiment through mathematical modeling techniques to seemingly objective algorithmic technologies. A common presumption within our burgeoning algorithmic culture is that mathematical discourse has the power to expose subjective bias and purge subjective judgment, yet those trained in rhetoric know that all symbol systems—whether mathematical or not—select and deflect reality, meaning that all symbol systems have horizons of judgment and that claims to objectivity often serve to conceal those horizons. Neo-Aristotelian scholars are hence inclined to examine not how math derives truth but how it enhances the power of arguments in the minds of particular audiences. How does the use of mathematics bolster an author's credibility? How can math create the appearance of objectivity? These are just some of the questions that a neo-Aristotelian approach would posit as useful starting points for critical engagement with algorithms.[23]

If our analysis stopped there, however, we would not get very far in revealing the horizons of judgment within complex algorithms. Those horizons are intricately woven networks of human and nonhuman agencies, assumptions, and delimitations that often begin in recognizably rhetorical forms of natural language, only to be translated into and transformed by mathematical discourse. To examine those practices of discursive constitution, one must think constitutively. A constitutive rhetorical approach invites scholars to engage with algorithms differently. As mentioned earlier, most critical scholars do not see a problem with algorithms as such, but instead with the people or institutions behind them.[24] From this perspective most if not all agency remains (comfortingly) in the hands of human actors, with algorithms as an extension or enhancement of that agency. A constitutive approach to symbolic action, however, sensitizes one to the *congealing of agency*, as human actors triangulate their thinking with nonhuman actors, including symbol systems, discursive formations, and environments of encounter. Such sensibility is alive to the agencies of discourse and symbol and how those agencies interact with and sometimes exceed human agency even as humans use those symbol systems to articulate "their" thoughts. One must be careful here, as it would be too convenient (both for governments and corporations) to offload agency onto algorithms, thereby minimizing individual and collective responsibility. Instead, a constitutive approach demands a symmetrical treatment of the various distributed agencies at play in complex algorithms. Analytically, this manifests as a tacking back and forth between traditional forms of human agency, the forces of agency within discursive formations, and the material agencies that algorithms both draw on and occasionally transform.

However promising this constitutive approach to algorithms, questions persist. If, as constitutive theorists claim, rhetoric is the material out of which new mathematical objects emerge (algorithms and otherwise), how exactly does that happen? Surely argument and inscription alone are necessary but insufficient conditions. Equally vexing, if algorithms really do reconstitute realities as they interact with those realities, how do we explain such alchemy? How, in short, does the language of math become materially manifest in the world?[25] To address these questions we need a translative rhetorical approach, which, as we know from our previous chapters, is designed to extend beyond symbolic-conceptual realms and study math as a practice of discursive-material assemblage. The task of a rhetorical study of mathematics shifts accordingly, focusing not on suasion in or through math but on how mathematical discourse translates the existing relations of a collective into new relations and how those new relations form novel hybrids of humans and nonhumans—and ultimately the consequences of those relations for the collectives we find ourselves in. This last step is crucial, for it is the tracing of an algorithm's materiality that will carry our analysis from the academic and technical spheres into the public sphere, where algorithms are rapidly reshaping material culture.

None of these approaches in isolation, however, would be able to examine the horizons of judgment within complex algorithms. Those horizons are diverse phenomena, full of intention, motive, analogy, delimiting assumptions, human-nonhuman hybrids, vertical and transversal relations, commensurabilities, and domains of validity, and that fails to mention how well these horizons are hidden from view, buried through strategies of argument and various appeals to realism (some subtle, some overt). This again is why I use the metaphor of *horizon*—something always far-off, present but intangible. Rather than take one approach or another, then, we need to draw from all our resources: from a neo-Aristotelian attention to argument structure and appeal, from a constitutive interest in symbolic action as the substance of mathematical invention, and from a translative focus on discursive practices of translation that actively transform problem-situations and create space for new hybrids to emerge. Only through this imbricated reading strategy can we begin to bring the horizon of judgment within the Li copula into view, see how it displaced practical judgment, trace the ways it translated and extended existing economic networks, and thereby understand it as an apparatus (like Archimedes's compound pulley or the Argand diagram) that mediated the recomposition of global finance.

The Li Copula as Principle of Composition

The term *copula* derives from the Latin word *copulare*, to connect or to join. By 2008 David X. Li's Gaussian copula had connected a great variety of financial entities, including most of the major investment firms, nearly all large banks, and even several key government rating agencies. He had, in the words of Felix Salmon, cracked "a notoriously tough nut—determining correlation, or how seemingly disparate events are related ... with a simple and elegant mathematical formula."[26] That algorithm, originally published in Li's "On Default Correlation" in 2000, quickly vaulted Li into the economic limelight. He became *the* master quant—a name used for financial engineers since the 1980s. He moved from the Canadian Imperial Bank of Commerce in Canada to a subsidiary of J. P. Morgan in New York to director of derivative investing at Citigroup, all in a few years. Many expected him to receive the Nobel for his work.

At a national financial engineering conference in 2003, he was regarded with esteem: "In front of a room of hundreds of fellow Quants ... he ran through his model—the Gaussian copula function—for default. The presentation," Sam Jones reports, "was a riot of equations, mathematical lemmas, arching curves and matrices of numbers. The questions afterwards were deferential, technical."[27] The Li copula was considered an ingenious breakthrough, a means of finally predicting with mathematical rigor the default risk between seemingly disconnected debt obligations (loans). His algorithm was used not only to determine default risk but also to rate and price that risk so that it could be sold (often as collateralized debt obligations, or CDOs) to government-backed financial institutions. The incredibly influential Alan Greenspan, who chaired the Federal Reserve from 1987 until 2006, championed these algorithmic technologies; convinced of their mathematical veracity, Greenspan claimed in 2005, "Recent regulatory reform, coupled with innovative technologies, has ... contributed to the development of a far more flexible, efficient, and hence resilient financial system."[28]

The vast majority in finance saw the Li copula as an effective means of predicting, pricing, and rating risk, and it was quickly integrated into the global financial system. In 2006 a mere six years after its publication, Stanford University professor Darrell Duffie noted that "the corporate CDO world relied almost exclusively on this copula-based correlation model."[29] And that "corporate CDO world" was growing exponentially—from $275 billion in 2000 to $4.7 trillion in 2006—accelerating toward a subprime mortgage crisis it could not foresee. How did this come to pass? Why did the global financial industry integrate the Li copula so quickly? How did that algorithm—a Gaussian *model* for *forecasting* default risk—become reified into an "innovative technology" of risk *measurement*?[30] And how did it eventually exceed the intentions of its architect and become a principle of composition within structured finance?

To begin to address these questions we must first understand the problem-situation that Li and the rest of quantitative finance faced at the end of the twentieth century. That story begins with a practice called "tranching," which arose in the 1970s as a way for large banks to create additional revenue streams. In short, tranching (a technique still practiced as of this writing) allows a bank to gather mortgages into a mortgage pool and then sell them as mortgage-backed securities (MBSs) to interested investors.[31] These MBSs are often tranched or, coming from the French, "sliced" into different categories of risk. Those categories typically range from AAA to Unrated, with many tranches (AA, A, BBB...) in between. AAA tranches are the least risky, offering the smallest return. Unrated tranches have the greatest risk and the greatest return.[32] Tranching enables banks to sell their mortgages but continue servicing them—thus continuing to profit while reducing risk exposure.[33] However, with tranching comes the problem of pricing risk: imagine you are an investor considering a purchase of an MBS. What is a fair return on investment for a AAA MBS or a BBB or an Unrated? For several decades this was a difficult problem within quantitative finance. Prior to Li's copula the techniques for pricing risk were considered approximate at best, in large part because no one could accurately predict default correlation between loans within a mortgage pool.

Why was default correlation such a difficult problem? Consider this simplified scenario: You own a house in Los Angeles. If you default on the mortgage, will that increase the chances of your neighbor defaulting? If so, by how much? What, in other words, is the default correlation between these two mortgages? Even more difficult, what is the default correlation between your mortgage and another random mortgage in Pennsylvania, New York, or any other state, for that matter? The answer is *it depends*—it depends on the relationship or lack thereof between homeowners, the local and national housing markets, and the national and global economies. Even this simple hypothetical, then, reveals the highly contextual nature of default correlation, which in mathematical terms means it is both unstable (possessing high degrees of variance) and extremely intractable (unable to be accurately modeled).[34]

The intractability of default correlation was a major problem within structured finance. Throughout the 1980s and 1990s, Wall Street quants deployed a number of strategies to approximate default correlation, including analysis of historical default data coupled with practical diversification strategies, but these were admittedly crude. As a result, rating agencies like Moody's and Fitch had an understandably conservative posture toward MBSs, granting high ratings to only the safest "senior" tranches. This in turn constrained the MBS market, since the major players Fannie Mae and Freddie Mac—as government-sponsored organizations—could purchase only "investment grade" securities (those with a

BBB-rating or higher). Within this environment banks could pool their mortgages and create tranched MBSs, but they could sell only the highest-rated tranches to major market investors, and, due to the uncertainty of default correlation, only a small portion of those tranches were given high ratings. The remainder of the MBSs were sold through what Joshua Coval, Jakub Jurek, and Erik Stafford call "'private-label' mortgage-backed securities"—a smaller market with less favorable terms for the banks.[35]

The problem with default correlation, in the end, was not risk but uncertainty. Investors don't mind risk, assuming they can reliably assess and price that risk; what they dislike is uncertainty. And default correlation was the poster child for uncertainty before the Li copula. Prior to that copula's emergence, circulation, assimilation, and eventual translation of global finance, investors used the heuristics of practical judgment to guide investment decisions. They examined the mortgage assets that backed MBSs and considered the relative diversity of those mortgages as a means of estimating the potential for default correlation, but these methods of judgment were themselves intractable—hard to control, human dependent, and time consuming. The Li copula rendered these forms of judgment (and their inefficiencies) superfluous, offering in their place what appeared to be a mathematically rigorous way to determine default correlation.[36]

The Analogy at the Heart of the Copula

How did the Li copula transform something as intractable as default correlation into an immutable mobile, an object perceived to be easily at hand and combinable at will, something that could transcend context and be used to make broad investment decisions (or, better yet, could—once the parameters of algorithmic automation were established—automate those decisions)? Understanding how default correlation was transformed into an immutable mobile requires a close examination of Li's article, where a fascinating rhetorical drama unfolds.

The first sentence of "On Default Correlation" is revealing: "The rapidly growing credit derivative market," it reads, "has created a new set of financial instruments which can be used to manage the most important dimensions of financial risk—credit risk."[37] Note how Li's discourse immediately constructs a risky and fast-moving scene—a reality that calls for action, rewarding those on the cutting edge of "new ... financial instruments." As testimony to this reality, an aggressive market already deploys various "instruments" for managing risk. These instruments are ubiquitous and they are, it goes without saying for Li's technical audience, mathematical in nature. At the outset, then, mathematical statements are taken up as objects—not formulas or equations or models but *instruments* that help weary investors navigate the "rapidly growing" markets. Revealing too is the presumed causal relationship between the market and financial instruments, the former calling forth the latter. Mathematical instruments, Li's discourse suggests, are deployed only out of necessity to cope with the increasing complexity of investment risk. This is how a realist drama unfolds, positioning Li not as advocate or interpreter but as unmediated witness to the realities of structured finance.[38]

Yet the list of "financial instruments" described in the first paragraph does more than establish a realist frame (a frame rhetorical scholars have shown to be the "default rhetoric" within economics).[39] It also builds a network, links Li's thinking to a variety of institutionalized financial products, and thus establishes those accepted financial practices as nodes of connection on which his work will build. "Credit default swaps," "equity tranche[s]," "collateralized loan obligations"—taken together, these signify an already established assemblage of financial instruments, products, and practices, all of which compose the credit derivative market even as their a priori existence reinforces the integrity of Li's realist style.[40]

As we know, this assemblage of "instruments" first emerged in the 1970s with the creation of mortgage-backed securities, at the same time—not by chance—that mathematization began to dominate the financial industry.[41] As we also know, the main constraint on growth was default correlation or, more precisely, how to accurately predict default correlation between loans. "Surprising though it may seem," Li's article attests, "the default correlation has not been well defined and understood in finance." Having established the epistemological problem, Li's article then critiques the existing solution—the "discrete default correlation" approach—as limited by time intervals when investors want a continuous distribution of default risk over time (five years, ten years, thirty years). This limitation calls for an alternative framework:

> Default is a time dependent event, and so is default correlation. Let us take the survival time of a human being as an example. The probability of dying within one year for a person aged 50 years today is about 0.6%, but the probability of dying for the same person within 50 years is almost a sure event. Similarly default correlation is a time dependent quantity.... This paper introduces a few techniques used in survival analysis. These techniques have been widely applied to other areas, such as life contingencies

in actuarial science and industry life testing in reliability studies, which are similar to the credit problems we encounter here.[42]

This passage radically transforms the problem called "default correlation" even as it offers a first glimpse of the copula's horizon of judgment. Notice that Li includes a seemingly innocuous example: "Let us take the survival time of a human being." This "example," however, foregrounds a trajectory of analogical reasoning that becomes clear only several lines later: if one considers mortgages analogous to people and default analogous to death, Li reasons, then using actuarial math from that domain becomes a potential means of analyzing default correlation.

This analogy marks the fulcrum on which the Li copula rests, but how exactly does it transform the problem-situation? Consider: prior to Li's work default correlation marked the boundary of computability in quantitative finance. If we accept Li's analogy, however, everything changes: actuarial science has used increasingly sophisticated algorithms to accurately predict life-expectancy correlations for decades. By analogy, if life expectancy correlations are predictable, then default correlation must also be predictable. The problem-situation transforms accordingly: default correlation no longer marks the boundary of computability; instead, it names a poorly understood but mathematically treatable phenomenon, which in turn transforms it from an intractable constraint on the credit derivatives market to a potential engine for its growth.

To better understand how that engine propelled financial markets forward, let's look at Li's analogy more closely, for it is no simple linguistic turn. Note, for instance, that if one accepts Li's analogy—that mortgages are like humans and default like death—the shift to survival science is quite logical. But that observation conceals the more significant transformation of mortgage defaults from *a scalable into a nonscalable phenomenon*. To elaborate: human longevity, at least currently, is a nonscalable phenomenon, meaning the largest instance is not much larger than the average. Average life expectancy might be 78 years, yet the chances of someone living even twice that long are nil (the record is 122). Mortgage default, in contrast, is scalable, meaning the largest instance can be many times the average.[43] Recall that in 2000 the main obstacle to predicting default correlation was the limited amount of historical default data that existed, indicating the average rate of default was historically low (which it was and is—well below 5 percent). Yet mortgage default rates can obviously double, triple, even quadruple the average. In 2004, for instance, the default rate on single-family mortgages was 1.39 percent; by 2010 it was 11.27 percent, or over eight times higher.[44]

Li's analogy, however, makes a multipronged argument: (1) that we have misunderstood mortgage default, (2) that it may not be scalable after all, and (3) that an overreliance on historical default data has distorted our perception.[45] Using historical default data, one can easily "prove" that default correlation is scalable. Yet that "proof" relies on scant data, possibly falsifying and certainly weakening any argument dependent on it. Li's analogy compels his readers to reconsider the value of historical default data: perhaps default correlation isn't scalable; perhaps historical default data is misleading. In so doing historical default data morphs from a valuable empirical resource for understanding default correlation into at best a constraint and at worst a distortion of the very phenomenon in question. But is there an alternative data resource to both clarify our understanding and improve the reliability of any predictive models of default correlation?

Enter credit default swaps (CDSs). The analogy between default and death granted Li access to all the mathematical tools of actuarial science, but without sufficient data no algorithm, no matter how sophisticated, could model default correlations accurately. Li needed more data, and his great innovation was to replace historical default data (whose ethos he had already undermined) with data from the CDS market. CDSs are the counterpart to traditional loans: with traditional loans a lender makes money off of loan interest, but with CDSs one offers insurance against default. If prices of CDSs increase, the probability of default increases. Li reasoned, if the CDS prices of two loans increased and decreased together, that indicated high correlation. This was the last major puzzle piece: Li had access to actuarial science and now he had the necessary data to build and test a copula.

Algorithms as Principles of Composition

We are now fully embedded in a hybrid—an admixture of the real and the imaginary, the actual and the virtual, the human and the nonhuman—a place where market risk is immanent, default correlation is poorly understood but calculable, mortgage defaults and death are akin, historical defaults are misleading, and credit default swaps are a rich resource of much-needed data. The problem-situation called "default correlation" has been deconstructed and many of its component parts transformed. Now it must be recomposed. For that task Li turns to mathematical algorithms as powerful principles of composition—techniques for rendering out of analogy and conjecture something hard, something tractable, something immutable.⁴⁶

But how do algorithms actively compose a world? How can we understand them as constitutive forces, ones that possess their own forms of agency? Perhaps a closer look at Li's mathematical discourse will offer some insight. As is common among mathematicians, Li begins with abstraction, inventing a "random variable called 'time-until-default'" and thereby transporting his reader from the empirical realm of the particular to the mathematical realm of the general.⁴⁷ This is a key rhetorical (read symbolically constructed) moment in most mathematical discourse, for it enables one to render commensurable what was initially incommensurable. In this case Li wants to treat all debt as commensurable, so he creates a *formal* category A to represent all loans, combining that with his "time-until-default" variable to form a new hybrid, T_A, which represents the time until default of any and all loans (now rendered equivalent through A). Li then represents the probability distribution (simply the points on a graph of the likelihood of default at different times) of A's default with $F(t) = \Pr(T_A \leq t), t \geq 0$ and the inverse, A's "survival" with $S(t) = 1 - F(t) = \Pr(T_A > t), t \geq 0$. Already apparent here is the way mathematical discourse enables one to "obtain nth order forms that are combined with other nth order forms coming from completely different regions," and those nth-order forms are part of the compositional power of algorithms.⁴⁸ In fact, it is the emphasis on abstract form that renders what was once intractably particular—the diversity of loans and their potential for default—into a commensurable and thus computable form.

Yet something more is afoot in these initial mathematical moves, for inside every mathematical code one can find the central ideas for which it is the vehicle. And the central ideas within the functions $F(t)$ and $S(t)$ are

1. that loans are like people and default like death;
2. that actuarial science can therefore be used to address default;
3. that default, through actuarial science, is best expressed as a probability function;
4. that, since default is nonscalable (just like human longevity), one can use Gaussian bell-curved math to effectively model default risk over time; and
5. that, as equivalent and commensurable probability functions, the individual particularities of each loan are unnecessary for assessing default risk.

In critically reading Li's mathematical discourse in this way a map of the copula's horizon of judgment begins to emerge, and, as that process evolves, we see how the judgments baked into that algorithm slowly displace the forms of practical judgment that guided financial decision-making prior to Li's work. Before the copula, for instance, practical judgment encouraged diversification as a heuristic to minimize default correlation.[49] That heuristic required evaluation of the particular debt obligations within a tranched mortgage-backed security. Li's actuarial approach, however, creates a world where particular mortgage defaults no longer exist—at least not in the same form. They have been transformed into commensurable probability curves, a transformation that promises to reveal the hidden patterns of default correlation while simultaneously rendering those old forms of *phronesis* (diversification, direct assessment) obsolete. Having set the foundations for his algorithmic treatment of default correlation, Li is now in a position to address more complicated default situations and complete his recomposition of the credit derivatives market.

With basic concepts and functions established for the default risk of a single loan, Li still needed to develop a mathematical way to express the more complicated problem of default correlation between loans. Predicting the default risk of a single loan is challenging but not impossible. Much more difficult are dependent or conditional probability problems, ones that ask, "If mortgage A defaults, what is the probability of mortgage B defaulting at time X." Here again, however, Li can rely on his original analogy, which sponsors multiple transversal connections between default and survival science. These transversal connections come in the form of algorithms designed to determine survival probabilities of one spouse, for example, if the other spouse dies. Transversally applied to default correlation, these more complex actuarial algorithms like the "hazard rate function," the "joint distribution function," and the "survival function" are combined and substituted into Li's original functions $F(t)$ and $S(t)$, such that we move from the original "survival function" $S(t) = 1 - F(t) = \Pr(T_A > t)$, $t \geq 0$ to its expression "in terms of the hazard rate function"

$$S(t) = e^{-\int_0^t h(s)ds}$$

to "the default correlation of two entities A and B":

$$\rho_{AB} = \frac{E(T_A T_B) - E(T_A)E(T_B)}{\sqrt{Var(T_A)Var(T_B)}}$$

One can easily become bewildered by these mathematical transformations, but one need not know the calculus of survival science to see that each step effects a slow scaling-up from the basic functions accounting for default risk of a single loan to the "comparison of groups of individuals" and their relative survival probabilities to "more complicated situations, such as where there is censoring or there are several types of default."[50] This process of scaling-up does two things: first, it strengthens the original simple functions $F(t)$ and $S(t)$ by association with other more complicated and well-established functions in survival science; and, second, each mathematical association and substitution improves the tractability of those original functions, increasing their ability to address more complex situations (mortgage pools, for example, or defaults between different forms of debt or even externally imposed censoring).

The remainder of Li's article leads the reader inexorably toward his ultimate algorithm. We are taken through his construction of credit curves based on CDS data, his analysis of other means of creating credit curves and their comparative weaknesses, and his introduction of mathematical copulas, all of which brings us to his final "bivariate normal copula function": $Pr[T_A < 1, T_B < 1] = \Phi_2(\Phi^{-1}(F_A(1)), \Phi^{-1}(F_B(1)), \gamma)$. The left side of this equation establishes a "joint default probability" function between loan A and loan B.[51] The right side of the equation then reexpresses that joint default probability as a copula function (Φ) that reduces default correlation to a single constant (γ). By the time we arrive at this algorithm, however, all the assumptions and delimitations with which Li's paper began are barely visible, obscured by the numerous mathematical steps taken—each of which are irrefutably "true" within the world the article constructs. The result is that, by the end of the article, the horizon of judgment and domain of validity informing the Li copula have almost completely vanished, buried in a series of *correct* mathematical steps.

This is how immutable mobiles are built. Through invention, substitution, and transversal connections an algorithm is assembled. That process of assembly transports us from the local (the singular mortgage) to the global (the universal debt-obligation-default-probability curve), with the ultimate purpose being to render the original algorithm into a universal computation machine. Why? So the resulting algorithm is powerful enough to function as a principle of composition—a "technology" that can gather, translate, reconfigure, and ultimately transform the social-material world. Mathematical algorithms are about not only solving problems but also enabling new possibilities—new hybrids—to emerge. Having traced the discursive practices used to assemble the copula, we can now turn toward those hybrids, attending to the copula's material recomposition of global finance.

Terraforming

The previous analysis examined the entanglement of traditional and mathematical strategies of argument within Li's article to show how those strategies translated and transformed the problem-situation called "default correlation," but how did the resulting algorithm materially impact global finance? Much is at stake in this question, for its address promises to reveal both the materiality of the Li copula and, more broadly, the terraforming modalities of complex algorithms.[52] By way of entry, let's begin with the vincula—the new relations that the Li copula made possible within structured finance.

Perhaps the most significant relation to emerge from the Li copula can simply be called the "algorithmic relation." Algorithmic relations are nothing new. When you use Google, Ways, Facebook, or Netflix you enter into various algorithmic relations. Yet in each case those algorithms are powerful for the same reason—they are able to interrupt a space of practical judgment and human agency and replace it with algorithmic automation.[53] You no longer need to engage your parahippocampus in the same way to navigate a city, nor do you need to read descriptions of particular films to find something to watch.[54] The algorithm's power lies in its capacity to displace (not eliminate but decenter and marginalize) traditional forms of agency and judgment. The Li copula is no different in this regard. It effectively interrupted a space of human agency and judgment, displaced it with a mathematical algorithm, expanded the realm of computability within quantitative finance, and provided a precise measurement of default correlation (within its domain of validity), conveniently reduced to a single constant. Evidence of its perceived veracity and tractability can be found in many locals, as it circulated quickly within quantitative finance. Li's status within that intellectual community grew rapidly, evidenced by both his climb up the corporate ladder and citation indexes within his field, yet the strongest evidence for the copula's acceptance was the fact that major rating agencies (Moody's, Fitch, etc.), whose job was to protect large institutions from excessive risk, adopted the copula as the primary means for determining default correlation.[55] Broad adoption of the Li copula allowed the algorithmic relation it introduced to spread, consolidating and displacing previous realms of human judgment while simultaneously expanding credit derivative markets in unprecedented ways.

The collateralized debt obligation (CDO) and its derivative the CDO^2 are perhaps the best examples of the Li copula's material recomposition of global finance.[56] CDOs are similar to mortgage-backed securities (MBSs) in that a securities firm pools and tranches debt obligations into structured investments; the main difference is that instead of pooling mortgages, CDOs draw from multiple sectors of the economy—everything from manufacturing debt to credit card debt. Invented in 1987, CDOs were thought to be relatively safe because they included diverse forms of debt from different economic sectors, but, just as with MBSs, the inability to accurately assess default correlation severely limited the CDO market. As a result, the CDO market was $69 billion in 2000, a mere fraction of the credit derivatives market at the time.[57] The Li copula, however, effectively erased those differences, using CDS data (which existed for most major forms of debt) to construct default probability curves and render all debt obligations commensurable. Suddenly, one could use Li's new algorithm to assess default correlation without attending to underlying collateral, for all loans were commensurable in the mathematical realm of the copula.

The Li copula, however, did not simply expand the fledgling CDO market; it enabled new products—new hybrids—to flood that market. The basic idea behind pooling and tranching, whether into an MBS, CDO, or anything else, is to create investment products that are safer than individual loans. Consider a simple two-loan pool. In this hypothetical pool each loan has a default risk of 10 percent, and those default risks are uncorrelated. If we pool and tranche those loans such that the junior tranche pays one dollar if neither loan defaults and pays nothing if either or both default, and the senior tranche pays one dollar in every scenario except if both loans default, then the risk of default for the senior tranche drops from 10 to 1 percent (since the junior tranche protects it).[58] Individually the loans would be rated as BB+ and thus could not be sold to government-sponsored institutions, but once they are pooled and tranched, the senior tranche would receive an AA-rating—well within the parameters of investment-grade bonds.[59] However, in our hypothetical we assumed no default correlation between our loans. If default correlation increases even marginally, we would get radically different risk exposures for our junior and senior tranches, which is precisely the problem the Li copula purported to solve. Armed with a mathematically precise measurement of default correlation, securities firms were able to create a variety of new financial products. They took those below investment-grade tranches from different CDOs and MBSs, pooled them together, tranched them, and, using the same logic, produced CDO^2s well above investment grade, something they could easily do because rating agencies used the same algorithm to rate the tranches as security firms did to create them.[60]

These hybrids of the Li copula and the algorithmic relation it introduced completely transformed global finance. As the CDO market expanded, the old division between mortgages and other forms of debt dissolved and the separation between the MBS market and the CDO market likewise dissolved. From 2000 to 2005 the CDO market grew from $69 billion to more than $500 billion, and by 2004 over half of the underlying collateral of CDOs were made up of mortgage-backed securities.[61] From 2003 until the financial crisis, the CDO market was the fastest-growing sector within structured finance; in 2006 alone sales for CDOs were $500 billion —as much as the whole CDO market the year before— and with increasing international interest the market grew to over $2 trillion, just shy of the GDP of China in the same year.[62]

These are not statistics—they are material formations, aggregates of the algorithmic relation introduced through the Li copula and circulated through its broad adoption by both sides of structured finance (the rating and investment sides). These numbers are traces of the terraforming power of the Li copula: the way it translated the problem of default correlation, shifted the border of computability, introduced an algorithmic relation that automated risk assessment, and enabled securities firms to do various forms of "ratings arbitrage," selling tranches initially below investment grade as AAA securities. Like Euler's famous equation $e^{\pi i} + 1 = 0$, the Li copula introduced a series of novel relations, out of which grew a host of new hybrids that Li himself could not have anticipated. Like the electrical revolution, those hybrids radically transformed power relations—in this case the relations between global finance, the government structures designed to regulate its many forms, and the citizens whom both are meant to serve.

Aftermath

We have already seen how the Li copula translated the problem-situation called "default correlation," introducing through a central analogy and a series of transversal connections a novel algorithmic relation that materially transformed structured finance on a global scale. Now we can examine the economic fallout, not to shift blame from humans to algorithms but instead to better understand the translative powers of mathematical algorithms (in this case within the context of structured finance). Only then will we be in a position to step back and consider the implications of this analysis more broadly.

Over a decade after the financial collapse of 2008, the anthropocentric narrative of greed, hubris, and ignorance that supposedly led to the Great Recession is well worn. In the *Financial Crisis Inquiry Report* of 2011 the authors conclude, "The crisis was the result of human action and inaction, not of Mother Nature or computer models gone haywire."[63] Michael Lewis's *The Big Short* blamed securities firms and rating agencies—the first creating "dishonest" investments and the second granting those investments inflated ratings due to "fat fees" from firms like Goldman Sachs.[64] Nobel laureate Paul Krugman targeted "politicians and government officials," who "should have realized that they were re-creating the kind of financial vulnerability that made the Great Depression possible"; Krugman characterized behavior before the financial crisis as "malign neglect."[65] Still others claimed that subprime mortgages spread due to perverse market incentives that rewarded banks for creating them.[66] Taken together these studies suggest no shortage of causes for the Great Recession. Yet, as divergent as they appear, they are also curiously similar: for in every account human agency controls the drama—whether in the form of greed, or ignorance, or "malign neglect"—with the ultimate purpose being to assign blame. They are as a result dangerously partial—anthropocentric in a way that obscures the power of algorithms as principles of composition that actively remake and expand our worlds.

What role, then, did the Li copula, our algorithmic exemplar, have in the economic collapse? In terms of implementation, it is easy to see how the copula enabled the spread of subprime mortgages. Equipped with a mathematically precise way to "determine" default correlation, the community we call "structured finance" no longer felt the need to investigate the underlying collateral of tranched securities. Combine that mentality with the hierarchical payout structure of tranching, and you have a recipe for the spread of subprime mortgages. Take all those low-rated MBS tranches, for instance, repool and retranch them, then rerate them using the Li copula, and you can create new highly rated securities. Your starting point for that retranching process, however, is radically different: instead of a pool with 5–10 percent subprime collateral (the historical average), you now start with 40–70 percent, yet you have a mathematical means to sell 70–80 percent of that pool as investment grade.[67] Practical judgment tells you that, eventually, those subprime mortgages must trickle up into the tranches sold to government-backed institutions. But with the dominance of the algorithmic relation—and the apotheosis of algorithms generally within modern finance—few actually looked at the underlying collateral, and that enabled subprime mortgages to infiltrate global financial networks at almost every level.

Even this narrative, however, remains too anthropocentric and instrumentalist to see algorithms as principles of composition. To escape the gravity of these paradigmatic narrative forms and begin to grasp mathematical discourse not merely as *reflective* of reality but also as *productive* of reality, we must consider the deeper structures within the Li copula. We can begin by returning to its central analogy and the difference between scalable and nonscalable systems. Recall that implicit in the analogy is an argument that mortgage default, like human longevity, is a nonscalable system. To support this claim Li used Gaussian probability theory and credit default swap (CDS) data to construct credit curves for probability of default over time, thus "demonstrating" the nonscalable nature of default. What went wrong? CDS data, while plentiful, was misleading. The CDS market was created in 1994, meaning the models were dependent on six years of economic data. During that time housing prices only increased, which artificially decreased default correlations between CDSs. Coval, Jurek, and Stafford's work shows that even a modest decrease in housing prices exploded the default correlations within the CDS market, suggesting that *when housing prices decrease default correlations increase in a scalable manner*.[68] The substitution of CDS data for historical default data thus introduces into the copula's horizon of judgment another assumption—that housing prices will increase—yet this assumption is curiously nonhuman, a feature of CDS data itself and not the consequence of the mathematician "behind" the copula.

Regardless of authorship (or its absence), the difference between scalable and nonscalable systems has serious consequences. Nonscalable systems can be effectively modeled using Gaussian probability theory (bell curves and normal distributions) because there is a built-in stability to these systems that keeps things "normally distributed"—that is, clustered around the mean with relatively few outliers. Gaussian models for *scalable* systems, however, are extremely error prone because, as Nassim Taleb points out, "the bell curve ignores large deviations . . . yet makes us confident that we have tamed uncertainty."[69] Unfortunately, the problem does not end there: Gaussian models not only predict scalable systems poorly but can actually conceal scalability even as they increase risk exposure. Outside of highly stable, normally distributed systems (such as life expectancy), correlation measures are fairly meaningless as a means of prediction.

The problem here is not with correlation per se but the reification of correlation into a "matter of fact," into something "hard" that one can use to make financial decisions. This is one of the roles that mathematical discourse plays in Li's article—it helps harden his analogical conjecture into an algorithm with apparent mathematical rigor. And within the world Li's analogy invents, his mathematics *is rigorous*. Yet outside that world—in the world of scalable financial markets and default correlations—the reifying rigor of Gaussian probability curves conceals risk for a simple reason: in normal-distribution probability theory there is an exponential decline in the odds of an outlier as one moves farther from the center. Put differently, when one uses Gaussian models to forecast, one implicitly accepts the assumption that as one deviates from the average the likelihood of an event becomes exponentially rarer, crashing faster and faster toward zero. When one develops mathematical models based on Gaussian mathematics, then, the judgment that the system modeled is "normal" (nonscalable) is already integrated into the mathematical structure, and the sophisticated mathematical models deployed effectively obscure the system's scalable risk.

Note the limits of both anthropocentric and instrumental narratives here: Li is not an all-powerful mathematical wizard, nor the unwitting architect of the subprime mortgage crisis. Instead he is a situated actor, and, when he speaks, whole discourses of power speak. He did not invent Gaussian probability theory, nor did he invent the quantification of finance movement that primed his intellectual community for algorithmic automation. These are broader discursive formations within which Li was caught and for which he was a spokesperson. He simply used widely accepted practices within quantitative finance and actuarial science to expand and test a thought-experiment. That process *brought him* to the Li copula. That copula is neither Li's invention nor his discovery; it is an apparatus that sliced the world of structured finance anew, introducing a novel relation, giving rise to myriad new hybrids, and transforming our economic structures accordingly.[70]

As an apparatus, the Li copula possesses its own unique ontological DNA, rearranging through various translations the hierarchy of evidentiary value, rendering that which was beyond the realm of calculation suddenly calculable, introducing new relations and novel hybrids, and thereby transforming the subjectivity of its users. As Finn observes, "Implementation runs both ways—every culture machine [algorithm] we build to interface with the embodied world of human materiality also reconfigures that embodied space, altering cognitive and cultural practices."[71] How did the Li copula alter cognitive and cultural practices? Beyond the commensurabilities it constructed and the dissociations it encouraged between investors and investments and beyond the algorithmic relation it introduced that enabled human actors to construct new human-nonhuman hybrids, consider the way the copula reduced default correlation to a single constant (γ). This feature of the copula, we must note at the outset, is not Li's—it is characteristic of mathematical copulas themselves. Yet that aggregation of default correlation into a constant had several cognitive and cultural consequences: it further reified the copula into a *technology* (instead of a model) for *measuring* (instead of estimating); it increased the perceived applicability of the copula across investment domains; and, most critically, it provided an anchor that promoted an image of stability, encouraging users to underestimate risk.[72]

While manifest in many contexts, the role of anchoring in "disjunctive events" underscores the copula's agential force. As Amos Tversky and Daniel Kahneman—who first studied the anchoring power of numbers—explain, disjunctive events take place within complex systems (such as nuclear reactors or human bodies) where many elements depend on one another for the system to function.[73] Within these systems, even though "the likelihood of failure in each component is slight, the probability of an overall failure can be high if many components are involved."[74] The network the Li copula helped produce had millions, perhaps billions, of components involved (CDS data, pooled debt obligations, Gaussian probability curves, CDO hybrids, securities firms, rating agencies, and so on), yet the risk of that complex network was gathered and condensed into a single number—a constant that belied the scalable risk of the new disjunctive system. That constant exacerbated the funnel-like structure of the Li copula, which homogenized the field of judgment that informed decision-making within structured finance and in so doing rendered that sector simultaneously larger and more fragile—that is, larger but at the same time increasingly funneled into a single node of judgment parameters that, if violated, would lead to cataclysmic collapse. In this way one can see that the copula did not simply "represent" a series of a priori relations within the market and distill them into mathematical form—it actively gathered the prior network and concentrated it such that it relied almost solely on the empirical veracity of CDS data. And because the CDS market was itself extremely scalable, the copula effectively increased structured finance's exposure to scalable change. As the copula was deployed, its anchoring effect diminished the perception of risk even as its funnel-like structure rendered the network called "structured finance" larger, more homogeneous, and, contra Greenspan, more unstable.[75]

Implications

The ways in which math can alter our cognitive and cultural practices are legion, yet to see these discursive-material entanglements we must extend our understanding of rhetoric, mathematics, and algorithms. Within the configuration that animates this analysis, for example, rhetoric expands beyond argumentation and symbolic action without leaving them behind and extends to include the study of symbolic-material relations, their emergence, their productivities, and their material consequences, adding to our accounts of words and deeds what Nathan Stormer describes as studies of "addressivity," or the ways symbolic-material practices establish a "set of capacities for address that forms and fades within fields of power."[76] One's scholarly positionality shifts accordingly, from the confines of negative critique—who did this, who's responsible, or even how did math betray us with obfuscation and reductionism?—to the realms of symbolic-material production, where we ask instead: How do symbolic-material practices make certain forms of addressivity possible? How does math as a symbolic-material practice translate, transform, and mediate, enabling novel algorithmic relations to emerge that expand our social collectives? What forms of agency must we account for to understand the unparalleled productivity of mathematical algorithms? And how can we trace the ontological force of algorithmic relations and the ways they shift our capacities for address—how they mark us as we mark them?[77]

These questions implicitly reject old modernist divisions between rhetoric and math, humans and nonhumans, societies and natures, not out of some antipathy for the past but because those ontological relations are becoming increasingly inadequate for understanding the proliferating realities of the twenty-first century. Instead, this chapter attempts to practice methodological symmetry, which requires one to attend to the multidirectional feedback loops between those divisions such that they begin to transform into hybrid networks, and we begin to see their productive and increasingly consequential relations. The exemplar in this chapter is the Li copula. As the analysis develops, we see how human and nonhuman agencies intertwine and congeal in the slow formation and articulation of the Li copula. We also see that the horizon of judgment within the Li copula is no simple distillation of human judgment (Li's or others') but instead an entanglement of human, mathematical, and nonhuman agencies. Just as crucial, the chapter demonstrates that a rhetorical approach to reading algorithms can trace the movement from the algorithm as symbolic apparatus to the algorithm as an agential culture machine that materially transforms social-material formations. One can see this process visually in the following diagram, which is a detailed extension of the diagram from chapter 2 (see fig. 10).

Algorithms are media of symbolic-material translation; they are discursive bridges that enable traffic between symbolicity and materiality. Their power emerges not merely from their promise of efficiency, consistency, or correctness but more critically from how they translate problem-situations, introduce novel relations into those problem-situations, and rapidly spread those new relations in materially transformative ways. Critical rhetorical analysis can trace this symbolic-discursive-material process of translation, revealing not only how

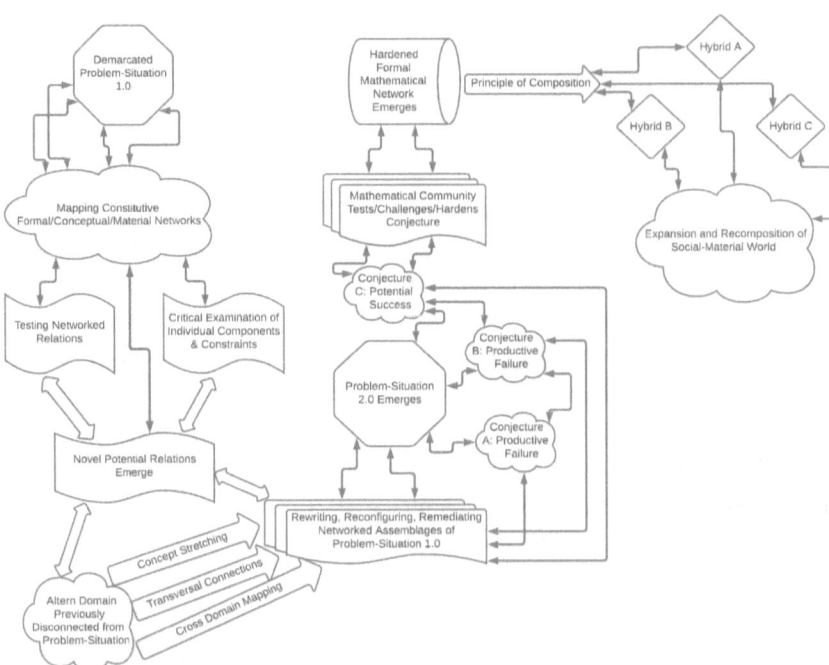

Fig. 10 | Microdiagram of translative rhetoric

economic algorithms are rhetorically constituted through argument and symbol but also how they activate their own hybrid agency through novel relations that on occasion radically transform the social collective.

Despite the eventual economic collapse of 2008, the purpose of this analysis is not to critique structured finance for being duped by an algorithm. The purpose is to offer an alternative to that epistemological (often positivist) critique, tracing instead the ways traditional rhetorical modalities (realism, analogy, association, dissociation) entwine with the rhetorical modalities of mathematics (abstraction, substitution, commensurability) to generate new relations between humans and nonhumans. Those new relations are the entities that mark the irreversible entanglement of symbolicity and materiality, the results of which often expand reality. This analytic approach seeks not only to reveal the ways algorithms can displace practical judgment. Perhaps more important, it seeks to promote critical engagement with practices of mathematization—algorithmic automation being one form—with the hope of encouraging others to become critical informants, citizen-scholars with the skills to unpack algorithmic black boxes within fields of power (whether economic or otherwise) and trace the ways they translate and expand our social-material worlds. For, as we will see in discussing the Anthropocene next, these translations do not merely expand social-material networks innocuously; instead, they can, and increasingly do, *threaten* the very social-material relations on which *Homo sapiens* depend.

6

Implications | Translative Rhetoric Revisited

We need to meet the universe halfway, to take responsibility for the role that we play in the world's differential becoming.
—Karen Barad, *Meeting the Universe Halfway*, 396

For thousands of years students of rhetoric largely ignored mathematics. Study of number and geometry might help sharpen the mind, Isocrates taught, but it did little to improve one's speeches, and its extensive pursuit took one away from more important matters of civic life and public affairs.[1] Of course, mathematics has always influenced public affairs, but prior to the twentieth century that influence was rarely felt on a daily basis at the level of the individual citizen. This is not to diminish mathematics' profound impact on everything from religion and music (Pythagoras, Pascal) to warfare and art (Archimedes, Da Vinci) to optics and physics (Newton, Leibniz) but merely to note that these influences could to some degree be separated from the everyday circulation of public discourse and production of public culture (the traditional domains of rhetoric). Only in the twentieth century did such separation become increasingly difficult to sustain, for it was in that century that one saw the rise of mass media, statistics and probability, computational power, surveillance technologies, and—perhaps the most immersive communication-tracking technology so far created—the internet. These are just a few of the phenomena necessary for the emergence of an information society, one in which those who can mathematically mine enormous data sets suddenly enjoy a power asymmetry heretofore unseen. Companies like Google and Facebook and Amazon place that power asymmetry in sharp relief, algorithmically mining and monetizing the digital shadows of every user (a revealing metaphor) that comes their way.

Math's influence on contemporary public culture, however, does not stop at the gates of Facebook's digital garden. Instead, mathematics has emerged in the twenty-first century as the metadiscourse of a rapidly expanding social-material world, a discourse so powerfully translative on everyday life—including on practices of communication, argumentation, and persuasion—that it can no longer be ignored if one hopes to offer a credible account of the rhetorical formations driving the production and circulation of public affairs. That is why I began this chapter with the final words of Karen Barad's wonderful book, *Meeting the Universe Halfway*, because seeing mathematics as a translative rhetorical force is one way that we can begin to take responsibility for the role we play in the world's differential becoming.

In chapter 1 we saw how Plato's thought pitted mathematics and rhetoric against each other, how the materiality of mathematical discourse was increasingly concealed from view as Plato's thinking progressed, and how a mythos of math emerged from his thought—one that makes taking responsibility for the role of mathematics in the world's differential becoming difficult if not impossible. This is because, for mathematical realists, mathematicians simply use mathematics to discover a priori truths about an ordered cosmos. Mathematics is a powerful language for representing those truths, but for mathematical realists it has no capacity to materially expand reality or accelerate its differential becoming. In this way Plato's mythos conceals in its unconcealment, math simultaneously emerging as a model against which to measure all other discursive practices (especially rhetoric) even as its own discursivity disappears from view. As a result of this perspective and its persistent dominance, we continue to have an impoverished understanding of how mathematical discourse evolves and how that evolution corresponds with math's increasing translative force on the world.

As good fortune would have it, a host of transdisciplinary scholars have already seen this problem and have begun to challenge the hegemony of mathematical realism and reject the oppositional dichotomy between math and rhetoric that Plato erected so long ago.[2] The earliest efforts to reconceive the relationship between rhetoric and math were radical in their time but fairly conventional in retrospect. By that I simply mean that fairly conventional Aristotelian theories of rhetoric tended to inform the earliest efforts to challenge their dissociation, which meant that, while rhetorical scholars might, for example, question the use of mathematics as an appeal to authority in public affairs, they had nothing to say about the formation of mathematical concepts themselves.[3] For that to come into play, for the materiality of mathematical discourse to emerge as a potential object of rhetorical inquiry, one would need a different theory of rhetoric, a theory capable of attending to both the role of structured forms of mathematical appeal, whether between mathematicians or laypeople, and the constitutive powers of symbolic action in the invention and dissemination of mathematical ideas. The notion of discourse and symbol as constitutive was much broader than the specific study of mathematical discourse, emerging in various fields simultaneously, yet one can see the hallmarks of that thought in the work of scholars like Rotman, Latour, and Lakoff and Núñez, whose work began to lay a basis for thinking of mathematical discourse from a constitutive perspective. One can also see those hallmarks in recent efforts by rhetorical scholars to understand the rhetorical materials out of which mathematical ideas emerge.[4]

With this book I have sought to build on the fine work of previous scholars and advance our understanding of mathematical discourse, how it has evolved over time, and what that evolution might teach us about math and its role in the world's differential becoming. While a traditional Aristotelian theory of rhetoric reveals how mathematics can be taken up and made to do the rhetorical work of persuasive appeal in public culture, and a constitutive theory of rhetoric examines the role of argument and symbolic action in the invention, formation, and dissemination of novel mathematical concepts, neither of those theories adequately accounts for the translative force of mathematics. By "translative force" I mean the ways math enables increased traffic between symbolicities and materialities and thus functions not merely as a reflection of the real but instead as an expansive multiplier of the real. To address these elements of mathematical discourse, I developed the notion of translative rhetoric, a theory of rhetoric designed to engage the rhetorical force of mathematical practice not at the level of argumentative appeal or at the level of invention but instead at the level of the vinculum (the relation). This, I argue, is where rhetoric and mathematics first meet and where we can often see the translative rhetorical force of mathematical discourse most clearly.

Although the theory of translative rhetoric is substantially different from either Aristotelian or constitutive approaches (see chapter 2), I do not see them as oppositional. Instead, I find they are most analytically powerful when placed cooperatively alongside one another, and the case studies of chapters 3 through 5 are organized to underscore that idea. Examination of the invention of the Calculus, for instance, combines an emphasis on the constitutive role of rhetoric in the formation of infinitesimals with the study of how seventeenth- and eighteenth-century intellectuals negotiated the infinitesimal's rhetorical force while arguing over its mathematical rigor. I then turn to the genealogical emergence of imaginaries, combining a constitutive emphasis on discursive invention with a translative focus on networks of mathematical relation, with imaginaries emerging as the consummate principle of composition and a powerful exemplar of how mathematical discourse fabricates and expands the real. In both cases we see how mathematical discourse simultaneously moved beyond the logics of representation realist epistemologies imposed on it and how that departure corresponded with the increasing translative force of mathematics on the social-material world. Finally, in the study of algorithmic culture and the subprime mortgage crisis, I make use of all three theoretical approaches to unpack the horizon of judgment buried deep in the Li copula and come to a better understanding of both that particular crisis and the potential ontological force of algorithms more generally.

Now, in these final pages, I would like to gather together as best as I'm able the overarching implications of this work for how we think about rhetoric, mathematics, and the ways they increasingly entangle with each other in contemporary culture. My hope is that, along the way, we will be able to distill (however temporarily) the potential translative force of the ideas at the heart of this book, which will help clarify the compatibilities and incompatibilities between this approach and more traditional notions of rhetoric and math as well as lead to some provocative discussions of potential openings for future research.

On Engines and Traffic

In the popular imaginary mathematics and rhetoric remain estranged, one an agent of infallible truth, the other an agent of manipulation and ideology. And while this estrangement might make sense as a rudimentary heuristic in everyday life (math often is true, after all, just as rhetoric is often manipulative), its reification into hard categories of difference comes at a high cost. For as we see in our study of Plato's dialogues (chapter 1), to purify mathematics of its earthly embodiments—both human and nonhuman—one must conceal the very practices of inscribing, arguing, and diagramming that constitute the rhetorical materials out of which new mathematics emerges and impresses itself on the world (humans included). Sans access to the materiality of mathematical discourse, one is then hard-pressed to explain its influence and growth, and so one invents fantastical metaphysical stories about a priori truth, the special relationship the human soul (or mind, or reason) has with that truth, and the inherent (often divine) order of the cosmos.[5]

While this perspective may have some psychological benefits—enabling those operating from within it to see their work as pure and absolute—it also has some substantial drawbacks. Mathematical realists, for example, tend to consider absolute truth the ultimate purpose of mathematics. Many scholars have described this as mathematical foundationalism (i.e., there is a deep mathematical foundation to the cosmos, and the mathematician's job is to discover it) and have linked it with an axiomatic approach in the classroom.[6] Within this framework, while there is an ineffable process of mathematical discovery personified by the "mathematical genius," mathematics as a field is primarily about formal proof, which establishes mathematical conjectures as irrefutably true. The purpose of pedagogy is, accordingly, to first teach students the many theorems and formulas that mathematicians have discovered and how to use those instrumentally to solve problems and, second, to teach students the proof procedures used to prove the infallibility of particular mathematical statements. This pedagogical approach is, of course, a major issue in the mathematics-education literature, where studies of student perspectives on math reveal two consistent themes: students perceive math as both abstract and rule-driven.[7] Mathematical realism strongly encourages such a perspective and, it has been shown, does not allow the majority of students to *identify* with mathematics.[8] One problem here among many is that, at the same time that mathematics increasingly shapes and constrains more and more of public affairs and human behavior, a smaller and smaller percentage of people possess the mathematical literacies necessary to understand that influence, much less interrupt it.

The analytic paradigm developed in this book calls for a radically different approach to both mathematics and rhetoric. To help guide an exploration of those differences, let's begin with an abbreviated list:

1. First, from a translative perspective the purpose of mathematics is innovation, and the consequence is often translation and transformation.
2. As such, the focus of mathematical practice and pedagogy should be distributed equally among the arts of innovation in informal mathematics and traditional practices of memorization, application, and proof procedures (these latter are still important skills but are no longer the primary focus).
3. A focus on the arts of mathematical innovation includes attention to mathematical practice and mathematical discourse, the ways those practices of discursive inscription translate the world into formalisms of commensurable in-*form*-ation, and how new principles of composition emerge from such translative practice.
4. Focusing on mathematical innovation reveals that strategies of rhetorical invention and argumentation (e.g., inscriptive translation, generalization,

analogical reasoning, metaphor, polysemy, notational invention, concept-stretching, and network building) are in fact engines of mathematical innovation and thus part of the evolution of mathematics itself.

5. Emphasis on mathematical practices of inscription, their evolution, and their translative force within historical contexts enables a paradigmatic shift from math as primarily epistemic (focused on discovery of mathematical truth and communication of said truth) to math as a reality-expanding, translative ontological force.

6. Underscoring the ontological force of mathematical discourse means rejecting a naive representationalist theory of language for a more comprehensive translative paradigm, one that understands the agential and diffractive force of mathematical discourse—how it takes up worldly relations and translates them into formal languages of commensurability that allow formal networks and novel potential relations to emerge, some of which become principles of composition that enable new decompositions and recompositions of the networks that compose the social-material world.

7. Attuning ourselves to the translative rhetorical force of mathematics implies a shift in not only how we understand mathematics but also how we understand rhetoric: whereas rhetoric traditionally begins with speech and the powers therein (communication, identification, argumentation, persuasion), here we must begin with inscription and rethink rhetoric in light of the powers of embodied inscriptive practice.

8. Rethinking rhetoric in terms of embodied inscriptive practice simultaneously expands the domain of rhetoric and makes room for specificity, meaning that rhetoric is not only "big" but diverse. That diversity demands not a "unified" metatheory but the development of complex critical vocabularies designed for inquiry into specific species of rhetorical practice.

9. Instead of speech or argument or audience or rhetorical situation, the object of study in this approach is the complex entanglement between relations (vincula) and inscriptions, and the goal is not merely to understand influence between humans but instead to understand how traffic between materialities and symbolicities happen and the role of that traffic in the world's differential becoming.

10. Shifting our focus from speech and argument to relation and inscription entails accounting for and addressing both human and nonhuman agencies as they congeal into novel hybrids with the translative capacities to slice the world anew; this will require a symmetrical balance between retrospective and prospective analysis.

11. Finally, attending to inscription and its translative material force brings one to a different understanding of matter and reality not as inert or stable "things" but as vast networks of phenomena with plasticity and vitality—networks whose relative stability we can no longer take for granted in the Anthropocene precisely because of the translative rhetorical force of our symbolic-material assemblages.

Let's develop each of these points in stages: First, I must say one last time that to describe math as rhetorical here does not mean it is another form of argument or persuasion. Instead, we are thinking of both rhetoric and math at the level of relations and the ways inscriptive practices enable the translation and assemblage of worldly relations into novel hybrids. Second, while mathematical truth remains integral to the practice and growth of mathematical knowledge within this approach (for truth as accuracy of logic, engagement and elimination of counterexamples, and so on are important criteria of judgment within mathematics communities), we must address the negative consequences of claims to absolute (T)ruth. These include (a) that any mathematical claim to absolute (T)ruth is unprovable and ultimately an article of faith; (b) that, even so, such claims to absolute (T)ruth often serve to reinforce a realist paradigm; and (c) that the realist paradigm encourages an unengaging deficit model of pedagogy, little insight into the arts of mathematical innovation, and a naive understanding of the powers of mathematical discourse (each of these points are developed at length in chapter 2). In contrast, a translative approach aligns much better with the actual practice of doing and thinking mathematically while simultaneously taking the agential force of mathematical discourse seriously.

On Vincula, Translative Rhetoric, and Rhetoric's Ontology

When one does mathematics, one often thinks relationally: relationships between numbers, geometric angles, equations, and sides of equations. As one advances in mathematics, one begins to think about broader relations between different branches of math such as geometry, algebra, and trigonometry but also about relations between mathematics and the material world. As an embodied practice of thinking and doing (as a performative practice), then, mathematics is deeply relational, deeply engaged with the vincula that compose the social-material world and the translation of those vincula into formal mathematics. That in fact is one of the sources of mathematics' ontological force: its capacities for translating worldly relations into ordered symbolic simulacra that can then be played with, reconfigured, and recombined into all manner of novel hybrids, some of which will have their own ontological force, their own powers of composition, and their own capacities as actants to introduce a novel relation into the network of relations composing the social-material world and accelerate the world's differential becoming in the process (by way of visual reminder, see fig. 11). To understand the liveliness of mathematical discourse, the ways it often emerges from and yet exceeds the bounds of human agency, and its increasing translative force in contemporary culture, I developed the notion of translative rhetoric.

Translative rhetoric begins with vincula—the relations that compose all phenomena—seeking to understand how those relations evolve and transform. It defines rhetoric as the embodied practice of inscription, the result of which is often translation and transformation. Thinking of rhetoric as embodied inscriptive

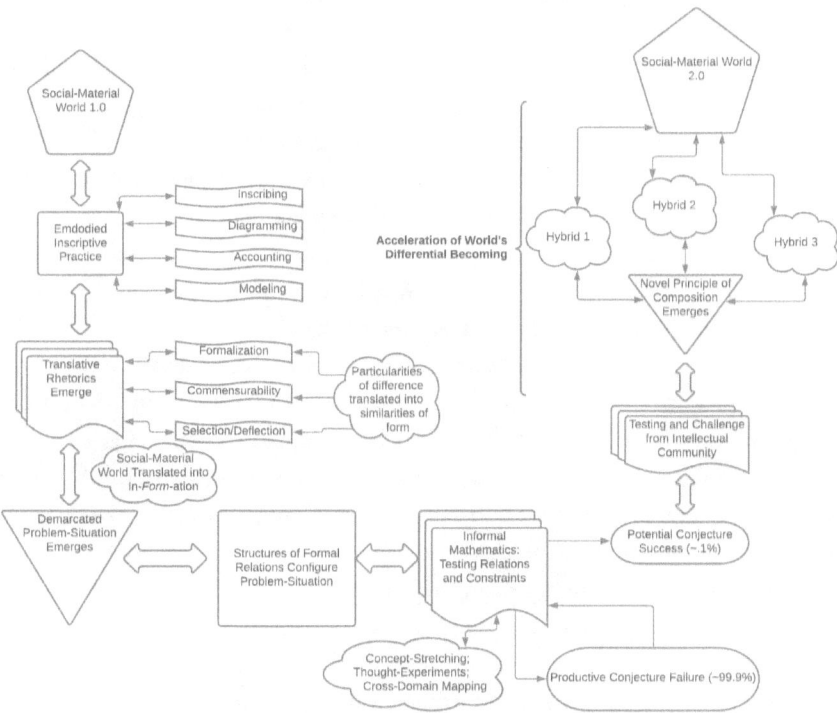

Fig. 11 | Diagrammatic model of translative rhetoric 2.0

practice has several ramifications: first and foremost, rhetoric here begins with inscription instead of speech. While such a difference might appear slight at first glance, it has some significant consequences. To begin with speech (traditionally conceived) is to structure rhetoric's ontology around and through the human, whose capacity for humanness is directly linked to the ability to transcend the modalities of brute force through speech and persuasion. To begin with inscription, in contrast, is to structure rhetoric's ontology around altern others—texts, diagrams, figures, sculpture, architecture, (photo-, tele-, helio-)graphs, computers, and so on—with which and through which we humans translate and transform the social-material world (including ourselves). It is to orient rhetorical study toward the "in" of inscription, which underscores the translative work of such practice, the ways inscriptions take up the radical heterogeneity of the world, translate it into new forms, and from that practice produce in-*form*-ation with its own translative rhetorical force, which is to say its own agential capacities to expand the very boundaries of what is thinkable and to open up space for new potential relations to emerge, some of which inevitably become novel principles of composition that radically rewrite and reshape culture. We saw this process take place in the emergence of infinitesimals and imaginary numbers, whose translative force lay in part in their capacities to socialize (to integrate into the networks composing the social-material world) both motion (rates of change) and recurrence and in so doing dramatically expand humanity's capacity to play with (compose, decompose, and recompose) those prehuman material phenomena.

The second major consequence of rhetoric as embodied inscription lies in challenging representationalist theories of symbolic action, for those theories render it difficult if not impossible to see the forms of translative rhetoric this book calls attention to in the first place. When one thinks representationally about discourse and symbol, one presumes those forms of inscription work through reflection, for "representationalism is the belief in the ontological distinction between representations and that which they purport to represent."[9] Such bifurcated thinking not only divides "the real" from "the word" (logos), concealing the material agency of inscription in the process, but also introduces "reflection of the real" as the ultimate purpose (telos) of inscriptive practice. For representationalists humans stand at the apex of a triangular relationship between the world (the known or knowable) and knowledge itself (see fig. 12). From within this configuration comes the amplified ontological value of the human as "knower" as well as various practices of transcendental epistemic purification deemed necessary for producing knowledge.[10]

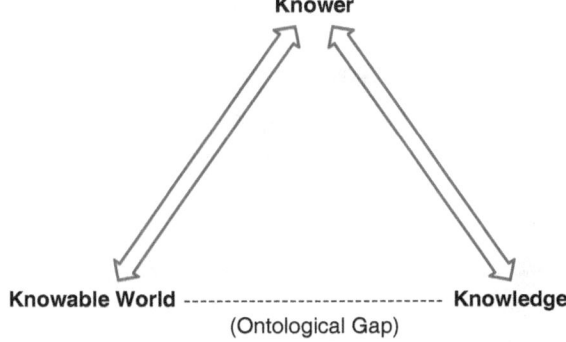

Fig. 12 | Episteme of representation

But if one wants to understand the materiality of discourse and the translative force of embodied inscriptive practice, one needs a different approach, which is why I suggest throughout the book that we think of the inscriptive practices of mathematics via the concepts of translation and diffraction instead of reflection. Translation and diffraction encourage one to see the practices of inscription as practices of translative rearticulation and remediation.[11] These practices do not mediate a relation between the separate realms of "nature" and "culture" but instead constantly produce and reproduce translative configurations of the social-material world. From this vantage point the awesome translative power of mathematics begins to make sense not by reference to an imagined a priori but by understanding how mathematical discourse and practice work as species of embodied inscription—as discourses (embodied inscriptive practices) that translate incommensurabilities into commensurabilities, fabricating in formal mathematical networks all manner of propositions, the great majority of which never see the light of day but for the rare mathematical proposition with the potential to become a new principle of composition, one with the translative force necessary to transform the relations of the networks that compose our world.

The final major consequence of rhetoric as embodied inscription is that rhetoric is no longer strictly human (or a phenomenon merely of, for, and between humans) but instead is situated and emergent in the liminal spaces between humans and nonhumans. For millennia the act of people persuading or dissuading one another first through the spoken word, then through the written word, and finally through various other media has defined rhetoric. This is not to say the field has been stagnant, for quite the opposite is the case, but rather that, generally speaking, rhetorical studies has orbited around the ontological act of humans using various media of expression to suasive ends. This paradigm for rhetoric is alive and well in the twenty-first century and deservedly so. We desperately need study and explanation of new forms of symbolic inducement emerging through social media networks, circulating through the internet, and finding their way onto the public stage, where politics—democratic and otherwise—have always played out. As a critical-theoretical orientation, translative rhetoric merely makes the case that another form of rhetorical action has been happening all the while, mostly behind the scenes and often deliberately hidden from view. That rhetorical action is happening at the level of relations and networks of relations, and it is not contained by the realm of human affairs. In fact, not only is this form of rhetoric not strictly human; it is also at least potentially completely nonhuman, since all sorts of animals and things have the capacity to translate the relations in which they are enmeshed (if this were not the case, for instance, there would be no such thing as evolution).

These shifts in rhetoric's ontology show that mathematical discourse and practice is as much an engine of rhetoric's evolution as rhetoric is an engine of math's evolution. For some time now rhetorical scholars have written about math and science, and the question often emerges: Are mathematical and scientific texts "*mere* rhetoric, constructing knowledge from socially inflected discursive practices, or are [they] . . . a reflection of some deeper reality over which rhetoric can make no claim?" This dichotomy, as Leah Ceccarelli points out, has functioned as a wedge in the "science wars," forcing what she calls a false choice between realism and relativism. In practice, however, rigorous rhetorical inquiry, for Ceccarelli, "must always strike a middle ground position between these two extremes," seeing the texts under scrutiny "as a convergence of discursive opportunities and material constraints." This nuanced position makes complete sense if one sees rhetoric as the study of textual argument. Yet that nuanced position does little to challenge the categories "realism" and "relativism," even if one posits a continuum of degrees between the two and imagines a rhetorical inquiry "designed to be sensitive to the recalcitrance of nature, just as it is designed to be sensitive to the possibilities of language."[12]

Shifting rhetoric's ontological focus to the translative force of embodied inscription, however, functions as an avenue of escape from both relativism and realism (something to which I imagine Ceccarelli is sympathetic), in part because, from the perspective of translative rhetoric, reality is not a stable, recalcitrant a priori. Reality is a complex network of relations with more plasticity than many feel comfortable admitting.[13] Mathematical, scientific, and technological innovation has made that plasticity more and more apparent as it accelerates up the hockey stick of Moore's Law, out of which has emerged a radically different rhetorical situation, one in which the stable relations of "reality" humans once took for granted—and the associated idea of the "recalcitrance of nature"—are increasingly being opened to decomposition, recomposition, and transformation. This new rhetorical situation has forced rhetorical scholars to go beyond traditional studies of arguments, audiences, and the texts that mediate between, and begin to consider what Lynda Walsh, Nathaniel Rivers, Jenny Rice, Laurie Gries, Jennifer Bay, Thomas Rickert, and Carolyn Miller have described as the "awesome flotsam of nonhumans" proliferating, influencing, and persuading in contemporary rhetorical culture.[14]

That engagement with nonhuman entities and the effort to account for them in our theories of rhetoric has had tremendous generative impact. We now have leading scholars in rhetorical studies developing case studies to account for the influence of code and computation (Carolyn Miller, Annette Vee, and James Brown); animals and ecologies (Diane Davis, Nathan Stormer, and Bridie McGreavy); and genetics and biomimesis (David Depew, John Lyne, Celeste Condit, and Roger Stahl)—all of whom find themselves pressed to develop new critical concepts to address their objects of study, and many of whom find traditional theories of rhetoric (designed to teach the arts of public discourse) wanting in terms of generating deep insight into the phenomena of interest.[15] Such research has culminated in an emergent paradigmatic shift in the field and the rise of what S. Scott Graham calls "rhetorical new materialism," an approach that seeks to diversify the humanist theories of traditional rhetoric with theories and concepts that can account for both humans and nonhumans in the production and circulation of public culture.[16]

While the theory of translative rhetoric developed in this book is tailored specifically to the study of mathematical discourse, it seems reasonable to suggest that such a theory and such a study might help ground the scholarly efforts of those interested in thinking about subjects like genetics or computation or automation from a rhetorical perspective. I use the word *ground* here first because understanding how mathematical discourses work—discourses that so often function as the bases on which new scientific practices and technologies assemble and activate their agencies—would significantly strengthen scholarly arguments about the rhetorical force of phenomena like computational code; genetics and humans' newfound capacities (through gene editing and nanobiotechnology) to decompose and recompose organic matter; or the rise of algorithmic economics that make decisions on the rather inhuman scale of microseconds. Understanding these newly emergent phenomena requires analysis of their fabrication, construction, and implementation phases. And while rhetorical scholars of science and technology have done much to understand the sociocultural consequences of implementation, we have not done as much to understand how the apparatuses of math and science are built, the constitutive networks of judgment therein, and the paths down which those networks lead. Part of that has to do with access and how technical mathematical discourse can function as an obstacle to inquiry. My hope is that the theory of translative rhetoric in this book, along with the critical case studies, can diminish our perception of mathematics as an obstacle and offer critical concepts for understanding its translative rhetorical force in a variety of contemporary contexts.

I say *ground*, second, because it seems the theoretical shift to embodied inscription both diversifies rhetoric's ontology and offers stronger footing for addressing from a rhetorical perspective the processes and consequences of mathematical and scientific invention—a topic of prime interest to rhetorical scholars of science and technology for at least the past fifty years. Over those decades we have thought of invention in math and science mostly through the traditional rhetorical lenses of argument; figurative language (analogy, metaphor, synechdoche); or framing practices and paradigmatic shifts (from which issues of commensurability and incommensurability emerge). And while these approaches have been incredibly productive, shifting our orientation just slightly to the translative powers of inscription seems to align in a different way with mathematical and scientific invention. For alongside and often antecedent to the macropractices of argument and persuasion either within or between technical and nontechnical communities, one finds micropractices of inscription, translation, and transformation, which not incidentally are what the broader macroarguments are often about.[17]

To underscore the difference here between a translative approach and a traditional rhetorical approach, consider the concept "incommensurability." From a Kuhnian perspective, incommensurability means epistemological incompatibility either within the history of a science (e.g., incommensurability between Ptolemaic and Copernican astronomies) or between scientific communities.[18] Rhetorical scholars, however, have challenged this idea of incommensurability, arguing instead that incommensurabilities between or within the sciences are as often "sociorhetorical" as they are epistemological. "Incommensurability," writes Carolyn Miller, "is an *impression*, which can be created not only by the differing intellectual commitments and habits that constitute a disciplinary matrix but also by argumentative positioning."[19] Here incommensurability is considered at the level of scientific communities and is figured as a rhetorical exigence. From a translative theoretical perspective, however, incommensurability is not just a feature of human epistemologies or communities; it is elemental to the complexity and heterogeneity of human and nonhuman ecologies and is central to the irreducibility of the phenomena and networks that compose those ecologies. Inscriptive practice often takes up that incommensurability and translates it into commensurabilities of form, from which in-*form*-ation and "data" emerge, simultaneously expanding the realms of calculability and increasing human capacities to intervene and reconfigure the networks of relations that make up the social-material world. In this sense commensurability and incommensurability are not about communities of humans or conflicting epistemologies but about heterogeneity, networks of relations, and forces of inscriptive translation.

In addition to these reasons, the analytic approach in this book might help ground rhetorical inquiry into math, science, and technology in a third way. Too often in the past, rhetorical criticism of math and science has fallen into the trappings of social constructivism, reducing these practices to the textual or symbolic and losing sight of the material relations these practices take up, translate, and transform.[20] Shifting one's orientation from speech and argument to inscription can help not only align rhetorical inquiry with mathematical and scientific practice—especially the practices of translation, invention, and assemblage—but also displace the critical impulse to "prove" the presence of rhetoric in math, science, and technology. The problem here is that rhetoric, traditionally defined as argument or persuasion, tends to signify to mathematicians and scientists the presence of fallibility (error) or subjective judgment or controversy, all things scientists and mathematicians seek to get beyond in their work. The resulting relationship between rhetoric and math or science is agonistic, and the impulse to prove rhetoric's presence is often seen as an attack on hard-won mathematical and scientific findings. With translative rhetoric, however, these relations transform considerably. Rhetoric, as conceived here, means embodied inscriptive practice instead of argument and persuasion. The purpose of rhetorical inquiry shifts accordingly, from studying how audiences are moved or how arguments constitute core mathematical and scientific ideas to understanding the translative force of embodied inscriptive practice. Put differently, the rhetorical inquiry advanced here calls for study of the translative force of embodied inscription within a world of elastic relations and in so doing perhaps offers a broader and less agonistic theoretical ground for the rhetorical study of math, science, and technology.[21]

This brings me to my final point about how translative rhetoric diffractively inflects rhetorical inquiry. So much of rhetorical scholarship in math, science, and technology has operated through the critical case study that there is a comparative absence (or at least asymmetry) of work seeking synthesis of those case studies into a more coherent whole and a sense that rhetorical inquiry into math, science, and technology both is limited to the symbolic and is predominantly retrospective.[22] Translative rhetoric addresses both of these challenges in different ways: first, it offers a means of potential synthesis both in its orientation toward embodied inscription and in its orientation toward mathematical discourse and practice. Second, in centering relations and the ways inscriptive practices translate and assemble networks of relations that can occasionally become principles of composition with material, ontological force, translative rhetoric further dissolves, and pushes rhetorical study beyond, the artificial modernist divisions between "nature" and "culture" or "materiality" and "symbolicity." Third, in centering its analysis around composition, decomposition, and recomposition, translative rhetoric seeks to balance its retrospective genealogical work with a *prospective* orientation—with thinking prospectively about the "horizons of judgment" within mathematical, scientific, and technological entities and where those horizons are likely to lead.[23] Concepts like principle of composition, discursive agency, and diffraction are designed to encourage prospective analysis, by which I mean a kind of analysis less interested in accounting for rhetorical "success" or "failure" and more interested in the transformations particular principles of composition portend, something it seems we desperately need as the implications of the Anthropocene come into view.

On the Anthropocene, Matter's Plasticity, and Rhetoric's Futures

The Anthropocene—a term coined in 2000 to describe the "geological time interval, in which many conditions and processes on Earth are profoundly altered by human impact"—implies a radically new rhetorical situation, one in which old ideas about humans, agency, matter, reality, and even causality must be substantially revised.[24] We have, for example, long conceived of humans as sui generis—distinguished in one way or another from other animals by the capacity for speech and reason, a special relationship with the divine, the capacity to reshape environments (homo faber), the power to think (cogito, ergo sum) and therefore act (as opposed to merely move or respond instinctually), the foreknowledge of death, consciousness, and so on. Despite their significant differences, however, every one of these arguments for the uniqueness of the human introduces an ontological hierarchy—a series of relations that amplify the ontological value of one (the human) while diminishing the ontological value of others (the nonhumans). And it is that ontological hierarchy that enables and justifies the objectification of our environments of encounter, transforming rivers into resources and animals into livestock and authorizing as legitimate the widespread commodifying practices of capitalism.[25]

Within these ontological hierarchies the uniqueness of the species emerges from the individual. Only in the individual human, biologically separate and independent from other humans, does one find speech, soul, spirit, mind, reason, thought, and will—each articulating a substantive ontological difference between the human and the nonhuman. Rhetoric here emerges as both a means to overcome biological division and further evidence of humanity's distinctiveness.[26] In the act of using speech to persuade, it has often been said, we *become human*—that is, more than mere beast. Through speech and writing, humans communicate their ideas and their perceptions of reality and generate alternatives to sheer brute force. Speech and writing, in this configuration, become tools for representing one's ideas, thoughts, and reality in general, from which come conventional notions of truth and knowledge as correct correspondence between representations and "the real." As Barad observes, "The representationalist belief in the power of words to mirror preexisting phenomena is the metaphysical substrate that supports social constructivist, as well as traditional realist, beliefs."[27] Within this episteme speech and writing become primary means of both activating one's own individual agency and producing knowledge.

It is difficult to fully understand the Anthropocene if one thinks representationally. From such a perspective, rhetoric (as symbolic action) may have the capacity to represent the real, but it cannot expand or transform the real. Reality is a relatively stable a priori—a hierarchically ordered ontology—something to be discovered, perhaps with the aid of rhetorical techne, but always already there nonetheless. Or, in contrast, if one is a social constructivist "the real" is still there, still a priori but simply beyond the reach of discourse and symbol. We do not ever apprehend the real through rhetorical techne but only produce various and competing social constructions of reality. Given these two options, one can see why most mathematicians are realists. But neither realism nor social constructivism can adequately explain the evolution of mathematics that has helped carry us into the Anthropocene. For, as we saw with the invention of both the Calculus and imaginary numbers, mathematical discourse has evolved in part through increasingly radical departures from the logics of representation that previously dominated mathematical thought. That departure has been infinitely productive, enabling the mathematization of rates of change and recurrence that, when combined, have dramatically expanded our understanding of the composition of several of the fundamental relations connecting the entities that make up our world. That understanding, of course, also enhances human capacities for decomposition and recomposition. And therein lies the reality-expanding power of mathematics—*not in the discovery of a priori relations but in the translative force of decomposition and recomposition that the mathematical socialization of a novel nonhuman relation enables.* It is those capacities for decomposition and recomposition that have transformed humans from a species made up of individuals gathered together for survival (much like other pack animals, with perhaps a few degrees of difference) into über-entities increasingly coming into conflict with other über-entities such as "ecology" and "climate."

The human as über-entity—that is, as an amalgam of human, discursive, and nonhuman agencies with enough geological force to define an age—is in part a product of math's translative rhetorical force.[28] Without the Calculus there would be no apprehension, prediction, or control of motion. Without imaginary numbers there would be no electrical revolution, comprehension of molecular structures, or understanding of the quantum, no awareness of Halley's comet, no industrialization, no combustion engine, no atomic bomb, no moon landing, no knowledge of DNA, no computers, no internet, and no cell phones.[29] Indeed, a great majority of the phenomena we consider as our objects and tools but that in truth have helped translate us into what we have become (hybrids, cyborgs, networks) would either cease to exist or would exist in such a radically different form that those phenomena—and those humans—would present to us as radically other. This is why one cannot understand the über-entity that is the human in the Anthropocene nor the Anthropocene itself sans an understanding of the translative rhetorical force of mathematical discourse.

The translative rhetorical force of mathematical discourse challenges us to rethink concepts like agency, human, and even matter itself, much less our notions of rhetoric and math. For one, we must accept that reality is not an a priori object, being, essence, or presence but instead a network of relations with varying levels of plasticity. The relations that compose our world and make what we call life possible are not immutable, unchanging laws but fortuitous relations fabricated in the geophysics of our universe. Some of those relations, like gravity, are so stable and essential to the composition of our world that they appear (even to Newton's great mind) as timeless law—absolute truth. Yet gravity too is a network of relations that might have been—in a different universe with a different beginning—otherwise. And while our engagements with gravity thus far suggest little plasticity, that does not mean we will never understand gravity in such a way so as to decompose and recompose it.[30] All matter, as Jane Bennett astutely noted, is vital—meaning both valuable and possessing vitality. This thought produces for Bennett "an ontological field without any unequivocal demarcations between human, animal, vegetable, or mineral. *All* forces and flows (materialities) are or can become lively, affective, and signaling. And so, an affective, speaking human body is not *radically* different from the affective, signaling nonhumans with which it coexists, hosts, enjoys, serves, consumes, produces, and competes."[31]

Phenomena with vitality are not beings—objects, things, commodities—but becoming networks of relations whose agencies have congealed into matter, some forms of which present to humans (for well-known evolutionary reasons) as independent objects in space, no matter how misleading that might in fact be.[32] This of course includes humans: we are vital matter, though not more vital than other forms of matter; we are also composite networks of the congealed agencies of humans and nonhumans. We might present as individuated independent bodies, but we know that to be illusion, that "our bodies" are in fact highly porous networks of teaming multitudes, with a universe of organisms residing in our stomachs alone. We also know that cultural practices and cultural institutions can and do materially mark, shape, and transform human bodies, as Anne Fausto-Sterling and Celeste Condit have shown with both practices of sexism and racism.[33]

What I hope this book has demonstrated is that mathematics too is vital matter; that mathematical discourse has its own vitality excessive to the intentions or agency of practicing mathematicians; that mathematical discourse, when combined with other human and nonhuman agencies, can fabricate (on occasion) principles of composition with enough agential force to radically reshape our social-material realities; that mathematical discourse has enabled the fabrication of novel hybrids like light bulbs, atomic bombs, or gene-editing machines that do not simply capitalize on the "secrets of nature" but instead *actively accelerate reality's becoming*—sometimes making amazing things like the understanding of molecular structure possible and sometimes enabling the destabilization of relations essential to particular ecological networks. In either case, however, a mathematical principle of composition materially expands the network of relations we call reality, often blurring and dissolving traditional demarcations between nature and culture in the process.[34]

In the diffusion of the distinction between "nature" and "culture," we are confronted by the imbroglio—that is, a reality composed of relations, some open to recomposition and others seemingly closed, but a reality nevertheless that is always diffractively becoming. Discourse, inscription, symbolic action—these are not mere practices of representation empty of ontological-material force; they are part and parcel of reality's becoming. And one of the prime questions for this book is to trace exactly *how* mathematical discourse inscribes itself on us, produces material bodies, and diffractively reconfigures the social-material world.

To understand that process within the mathematical context, however, we have to rethink rhetoric itself—not as speech or argument or persuasion but as the study of vincula and the translative force of inscriptive practice. For this book, for understanding the movement between mathematical discourse and the social-material world as well as the consequences of said movement, inscription and vincula are our foci of study. Mathematics is a powerful means of configuring and reconfiguring relations, but mathematicians are not generally interested in the materiality of their discourse. Mathematicians want to do mathematics. Yet if *we* want to understand how mathematical discourse becomes materially manifest, increasingly shapes public culture, and impacts real bodies and spaces, we need an approach that can trace its translative force. That is what the theory of translative rhetoric in this book is meant to do.

I do not claim, to be clear, that I have discovered some deeper essence of rhetoric. For I agree with the many rhetorical scholars who warn against treating all rhetoric as if of one realm, as if it might be a singular entity or a unified whole (however "big").[35] There is a deep impulse within our field (both objectivist and modernist, often unconscious) to think of rhetoric as an independent entity, whence comes the concomitant desire to "know" it, to learn of its various dimensions, to discover its surfaces and depths, to reveal its secrets. Such an approach, however, seems to implicitly ignore or reject rhetoric's vitality—that is, its capacity to evolve and diversify and multiply and become something other than it was in 500 BCE (or any other arbitrary point in the past). A more honest—and, to my mind, more interesting—approach embraces a diversity of rhetorical ecologies and works to tailor one's network of concepts to the study of one's particular rhetorical domain, something many rhetorical scholars are already doing.[36] This is not to say that a concept from one domain might not be applicable in another, but instead that when one *stretches* a concept from one domain to another (from the oratorical to the mathematical, for instance) one will diffractively produce, if not a new concept, certainly a new meaning for the concept involved.[37] We err in pretending otherwise, not only because different rhetorics thrive in different domains but also because in disregarding domain specificity we undermine the case for the ontological-material significance of rhetorical practice.[38]

I contend that, as a discursive ecology, mathematics renders most vivid the ontological-material significance of inscription, of the potential of symbolic statements to become principles of composition with the translative force to remake our social-material worlds. And it is that increasing translative force that defines what Freeman Dyson calls "cultural evolution." Most know some version, however partial, of the story of biological evolution, but only recently have we come to fully appreciate the influence of cultural evolution. Dyson draws the distinction thus: biological evolution is about the "spread of genes" while cultural evolution is about "changes in the life of our planet caused by the spread of ideas."[39] For Dyson, prior to the agricultural revolution, biology was the prime mover of evolution. After the agrarian shift culture became increasingly influential. Yet, given the arguments made here, this formulation smacks too much of the old divide between nature and culture. Instead, we might offer a friendly revision to Dyson's account: both biological and cultural evolution are about changes in the life of our planet caused by the spread of relations; biological evolution (as far as we understand it) happens via mutation, gene flow, genetic drift, and natural selection (driven in part by a species' fitness to a particular environment). Cultural evolution happens in part via the fabrication of principles of composition, which enable the creation of novel hybrids that recompose existing social-material relations (see fig. 11).[40] Those novel hybrids can enable fantastic forms of understanding, but they can also homogenize diverse fields of power. And while homogeneity might be useful for algorithmic automation, it is often anathema to both social stability (as we see with algorithms and the 2008 financial crisis) and ecological vitality. This to my mind is what careful study of mathematical discourse can reveal—not only how mathematics evolves and grows but how it functions as a primary discursive means of translation of both biological and cultural vincula, enfolding the two into increasingly complex networks that have upended binaries between nature and culture as well as human and machine, completely transforming the social-material world and enabling mathematics to become the metadiscourse of biocultural formation (here "cultural" includes the forces of technological evolution).

The world of entangled biocultural formations is the world of the Anthropocene.[41] It is our world, and we must produce rhetorics equal to it. We have answered the question of basic human survival within a harsh and indifferent environment. The Anthropocene poses a different question: Will we survive our own growing capacities for decomposition and recomposition of the social-material world? Will we newly emergent über-entities called humans continue to operate as if we are transcendental beings endowed with the powers of composition, ontologically superior, dissociated from the ontologically less (nonhuman), and willfully ignorant of the ecological and cultural consequences? Or will we come to understand ourselves and our environs not as two worlds (whether nature/culture, science/society, human/nonhuman, etc.) but as a singular "knotted world of vibrant matter," one in which we collectively come to see that it is in the self-interest of the species to understand how our practices of decomposition and recomposition can and often do destabilize the very relations that make humanity's survival possible?[42] These are just some of the questions the future poses to both our field and humanity in general. And if this book has any translative force of its own, which I hope it does, it will have raised many more questions than it has answered. What do we do now that we are alive to the translative force of mathematical discourse, for instance? What might a fully developed prospective form of rhetorical inquiry look like? How can we keep up with the rapidly accelerating translative force of emergent principles of composition? Can we continue to maintain a deliberative democratic culture in the face of such acceleration, or must we imagine new structures of social-material formation responsive to new social-material relations? All of these and more confront us in the Anthropocene. I wrote this book to help us see more clearly just how mathematical discourse has evolved and, in the process, how it has helped write the Anthropocene into existence. For if we can write ourselves into the Anthropocene, perhaps we can write ourselves beyond it as well.

Notes

Introduction

1. Wasserman, "Hating Gerrymandering."
2. See, as starting points, S. Kennedy, "Electoral Integrity"; and Issacharoff, "Gerrymandering and Political Cartels."
3. Hill, "Wrongfully Accused." See also Moravec, "Do Algorithms Have a Place."
4. The US Technology Policy Committee concludes, "when rigorously evaluated, [facial-recognition] technology too often produces results demonstrating clear bias based on ethnic, racial, gender, and other human characteristics recognizable by computer systems. The consequences of such bias, USTPC notes, frequently can and do extend well beyond inconvenience to profound injury, particularly to the lives, livelihoods and fundamental rights of individuals in specific demographic groups, including some of the most vulnerable populations in our society" ("Statement on Principles").
5. Cyranoski, "What CRISPR-Baby."
6. Ledford, "Quest to Use CRISPR." For more on the technical history of CRISPR, see Ishino, Krupovic, and Forterre, "History of CRISPR-Cas."
7. I unpack the concept of vincula more extensively in chapter 2, but in short it names the relations that compose the networks that form the social-material world. Throughout the book I argue for a translative theory of rhetoric that makes vincula (instead of persuasion or argument or symbolic action) the foci of study because it is in the production and circulation of relations that rhetoric and mathematics meet.
8. Prior to the twentieth century, few questioned these fundamental metaphysical distinctions. Are there a few rare exceptions—exceptional intellectuals—who challenged these binaries in one way or another? Of course, but not until recently has such thinking become more mainstream (at least within academic circles) and increasingly accepted as reflective of the lived experience of individuals in the twenty-first century.
9. See Dyson, "Biological and Cultural Evolution."
10. The notion of scientific or mathematical black boxes is widespread across multiple literatures, from science studies to critical algorithm studies, and refers to the ways complicated scientific and mathematical discourse can render the ideas within largely inaccessible to non-expert audiences.
11. Asimov, foreword to Boyer and Merzbach, *History of Mathematics*.
12. I borrow this notion of vincula from Bruno Latour, who develops it throughout his many projects (see especially *Pasteurization of France* and *We Have Never Been Modern*). However, throughout the book the concept takes on additional meaning and texture from its use in mathematics (as a signifier of the union of several mathematical figures) and other discursive fields such as anatomy (as connective tissue). By "principle of composition" I do not mean a principle of writing in a composition classroom but instead a principle through which the networks of relations that compose our world are fashioned. See chapter 2 of this volume for a full treatment of this idea.

13. There are many books about the abuse of mathematics for political or ideological ends. One of the best remains Gould, *Mismeasure of Man*. For more recent cases, see O'Neil, *Weapons of Math Destruction*.

14. Aristotle, *Rhetoric* 1.2, 1355a.

15. On the division between nature and culture as sometimes implicit and sometimes explicit in social constructivism, see Foucault, *Order of Things*; on the manifestation of this division in science studies, see Latour, *Science in Action*; Latour, *Pasteurization of France*; Latour, *Pandora's Hope*; and Barad, *Meeting the Universe Halfway*.

16. I develop this difference and its connections with research in rhetorical studies extensively in chapter 6. The best case I am aware of for rhetoric as diverse (or "polythetic") comes from Stormer's recent work; see "Rhetoric's Diverse Materiality"; Stormer, "Articulation"; and Stormer and McGreavy, "Thinking Ecologically."

17. For a classic example of the rhetorician's critique of a mathematical-realist philosophy as applied to the public sphere, see Perelman and Olbrechts-Tyteca, *New Rhetoric*, as well as Perelman, *Realm of Rhetoric*.

18. Lakatos, *Proofs and Refutations*, 1–2.

19. Davis and Hersh, "Rhetoric and Mathematics."

20. Thurston, "On Proof and Progress," 162.

21. Cifoletti, "From Valla to Viète"; Cifoletti, "Mathematics and Rhetoric."

22. M. Ascher, *Ethnomathematics*; Richland, Holyoak, and Stigler, "Analogy Use"; Ernest, *Social Constructivism*.

23. For analyses of the historical forces that have long alienated rhetoric and mathematics as well as the various contemporary scholarly efforts to challenge that alienation, see Wynn and Reyes, *Arguing with Numbers*, especially chapters 1 and 2. For more on the rise of new materialist theories in rhetorical studies, see Graham, *Where's the Rhetoric*.

24. G. Kennedy, "Hoot in the Dark," 2.

25. Gaonkar, "Idea of Rhetoric." The original essay was revised and expanded in Gross and Keith, *Rhetorical Hermeneutics*. For those interested, Gross and Keith's *Rhetorical Hermeneutics* is an excellent collection of essays responding to Gaonkar's argument.

26. Gross, "ARST Oral History Project."

27. In Gross's words, "What you have is [rhetorical scholars] who are not able to carry out a full program because they're . . . dealing with tools that are inadequate to the job. That's basically it. And I think that's the problem with the people who are contributing to the rhetoric of science now; that's why they deal with public policy. It's easier with public policy since public policy is not science" (ibid., 10:45–13:42). See also Gross, "Alan Gross in His Own Words," and Gross, *Starring the Text*.

28. Gaonkar, "Idea of Rhetoric" 29.

29. On this Aristotelian distinction and how new materialist theories challenge it, see Prenosil, "Bruno Latour Is a Rhetorician."

30. The list of publications for this work is extensive, but a few touchstones might include Vivian, *Being Made Strange*; Stormer, "Rhetoric's Diverse Materiality"; Stormer and McGreavy, "Thinking Ecologically"; Davis, "Rhetoricity at the End"; Davis and Ballif, "Introduction"; Benoit-Barné, "Socio-technical Deliberation"; Greene, "Rhetoric and Capitalism"; Biesecker and Lucaites, *Rhetoric, Materiality, and Politics*; and Rickert, *Ambient Rhetoric*.

31. Davis and Ballif, "Introduction," 347.

32. A few starting points for each of these: on animal and other nonhuman rhetorics, see Davis, "Rhetoricity at the End"; on scientific discourse as productive of novel hybrids, see Lynch and Rivers, *Thinking with Bruno Latour*; and on posthumanist ecological rhetoric, see Stormer and McGreavy, "Thinking Ecologically."

33. One reason for realism's persistence is the experience one has when doing math of engaging in or with something larger than the self, as well as the sense of constraint that mathematical symbolics often impose. Perhaps because we lack a better alternative, we often anthropomorphize mathematics, rendering it a sign of transcendental being. From a rhetorical perspective, however, these experiential phenomena are more evidence of the agential power of mathematical discourse. In the mathematical worlds one creates with math, for instance, one's actions are highly constrained by the logics and structures of the mathematical discourses one *thinks through*. The many ways that mathematical discourse shapes and constrains the imagined worlds within which mathematicians think underscores the agency of mathematical discourse—its capacity to surprise, to be an interlocutor, and to resist conjectures that violate its logics.

34. As Davis and Ballif put it, "Traditionally, rhetorical theory has been defined as the study of human symbol use, which posits at the center of 'the rhetorical situation' a knowing subject who understands himself (traditionally, it is a *he*), his audience, and what he means to communicate; indeed, this capacity to mean what he says and say what he means is, putatively, what distinguishes him *as* human" ("Introduction," 347). This differentiation between action and motion is fundamental to Burke's theory of rhetoric, which has been incredibly influential within the field. See *Grammar of Motives*.

35. This experience of interlocution and surprise has led most mathematicians to mathematical realism as a straightforward and politically advantageous explanation of the power of mathematics. Those advantages, however, come at the cost of a deep and broad understanding of how mathematical discourse works and how it increasingly shapes public culture.

36. The Calculus and imaginary numbers were also crucial discursive apprehensions of two fundamental phenomena of our social-material world: rates of change for the Calculus, and recurrence for imaginaries, and it is from their combination that humans have harnessed everything from steam power to electricity to digital information transmission.

37. Barad's work on diffractive analysis is second to none and significantly influenced the approach to mathematical discourse and practice found in this book. See *Meeting the Universe Halfway*.

38. Schiappa, "In What Ways," 34.

39. Wynn and Reyes, "From Division to Multiplication," 28.

40. Schiappa, "In What Ways," 38.

Chapter 1

1. See Fowler, *Mathematics of Plato's Academy*, 200–201, which refers, for the sources of the inscription, to an article by Saffrey, "Ageômetrêtos mêdeis eisitô," reprinted in Saffrey, *Recherches sur le néoplatonisme*.

2. Boyer and Merzbach, *History of Mathematics*, 101.

3. In an essay challenging Platonic realism, Darek Abbott estimates that approximately 80 percent of mathematicians are realists ("Reasonable Ineffectiveness").

4. Hacking, "Why Is There Philosophy," 3; Changeux and Connes, *Conversations on Mind*, 12. See also Hacking, *Philosophy of Mathematics*. To be fair to Connes, he proposes a more modest form of Platonism in which there is some room for human invention, but there is still a primordial mathematical reality beneath those inventions and on rare occasions a mathematician discovers a new aspect of that primordial reality.

5. See Gödel, *Collected Works*.

6. Einstein, *Sidelights on Relativity*, 27.

7. Rotman, *Mathematics as Sign*, 29–30.
8. Fowler, *Mathematics of Plato's Academy*, 7. Fowler also dates it "from about 385 BC" (7). All citations of *Meno* come from Plato, *Laches, Protagoras, Meno, Euthydemus*.
9. The point about purification of discursive content becomes important later, when distinguishing between Plato's views of mathematics and a rhetorical view. This is not to say that Greek geometers never used letters in their diagrams but rather that the practice of lettering, for Plato, played only a pedagogical or communicative role and was ultimately superfluous to geometric demonstration itself.
10. Aristotle, *Posterior Analytics* 75b15.
11. For a more extended analysis of this sense of space for the Greeks, see Fowler, *Mathematics of Plato's Academy*, 7–14.
12. Arndt and Haenel, *Pi Unleashed*, 175, 205.
13. See Plato, *Republic*, bk. 7.
14. As Vlastos notes, "It would be reasonable to allow for substantial development between the *Meno* (where the new theory of forms has not yet been formulated) and the *Phaedo* (where it is presented explicitly as the foundation of Plato's metaphysics." "Elenchus and Mathematics," 382.
15. When Plato wrote *Meno*, these ideas were still in their developmental stage; they are more clearly articulated in the middle books of *The Republic*, especially books 6 and 7, where his negative views on the practical applications of mathematics through calculation become clear, as does his apotheosis of geometry.
16. See Vlastos, "Elenchus and Mathematics," 386–87.
17. For more on scholarly resistance to Plato's Pythagoreanism, see J. Kennedy, "Plato's Forms," 20.
18. I use the plural "philosophies" here because in Plato's time there were already many different views, some of them contradictory, held by those called Pythagoreans. A comprehensive review of the literature can be found in Horky, *Plato and Pythagoreanism*. For more on Pythagorean philosophy, see Burkert, *Lore and Science*; Riedweg, *Pythagoras*; and Zhmud, *Pythagoras and Early Pythagoreanism*. For an explication of how the study of geometry influenced Plato's thought beyond the sources cited here, see Vlastos, "Elenchus and Mathematics."
19. J. Kennedy, "Plato's Forms," 2–3.
20. See Heath, *History of Greek Mathematics*, vol. 1.
21. It is striking to see these views alive and well in so much of contemporary popular culture; see, for example, Tyson's acclaimed 2014 television series *Cosmos*. For more scholarship on Pythagoras and the Pythagoreans, see Guthrie, *History of Greek Philosophy*; and O'Meara, *Pythagoras Revived*.
22. J. Kennedy, "Plato's Forms," 17.
23. See Horky, *Plato and Pythagoreanism*, 190. It's fascinating to find that the debate over the fallibility or infallibility of mathematics and thus its ultimate ontological status is as old as the practice of thinking mathematically.
24. Note that the form of this dialogue is different from *Meno* partly because the events that took place in Socrates's cell are being recounted by Phaedo to Echecrates and partly as a sign of Plato's own intellectual movement away from the Socratic method of *elenchus* (see Vlastos, "Elenchus and Mathematics"). I have purposely allowed the form of the dialogues I cite to evolve to reflect Plato's own evolution, as can be seen here and even more so in citations of the *Republic* to follow.
25. Vlastos makes a similar argument: "When Plato's epistemology has matured, as it will by the time he comes to write the middle books of the *Republic*, he will be qualifying this first starry-eyed view of geometry, insisting that the axioms of geometry are not the first principles which unphilosophical mathematicians take them to be: they should be regarded as

'hypotheses' which are themselves in need of justification; mathematicians who treat them as final truths are only 'dreaming about reality' (533b–c). In the *Meno*, no such caveats are even hinted at" ("Elenchus and Mathematics," 376)

26. A close reader of Plato will see the number three playing a vital role in many of his dialogues—the tripartite division of the soul in *Phaedrus*, the trivium of mathematical study in the *Republic*, the three sections of the divided line in the *Republic*, the counting out of the triad of interlocutors to begin *Timaeus*, and so on.

27. Of this ambiguity regarding Form-Numbers and their relative location vis-à-vis the realm of ideas and the realm of appearances, see Horky, *Plato and Pythagoreanism*, 188–99.

28. In the dialogue Socrates says, "The image must not by any means reproduce all the qualities of that which it imitates.... Would there be two things, Cratylus and the image of Cratylus, if some god should not merely imitate your color and form ... but should also make all the inner parts like yours ... and in short, should place beside you a duplicate of all your qualities? Would there be in such an event Cratylus and an image of Cratylus, or two Cratyluses?" (Plato, *Cratylus* 432b–c).

29. Horky, *Plato and Pythagoreanism*, 163.

30. For a detailed analysis of the practices of ancient Greek geometers, see Netz, *Shaping of Deduction*.

31. Ibid., 66, 62–67.

32. Plato, *Plato*.

33. Nussbaum, *Fragility of Goodness*, 110.

34. The desire for a science of judgment is alive and well in contemporary culture. Chapter 5 is in large portion about the desire for a science of judgment in contemporary economics and the consequences of that desire as they were made manifest in the 2008 financial crisis.

35. For additional evidence beyond *Gorgias* and *Phaedrus* that Plato considered geometric demonstration as a model for approaching problems and making arguments, see Plato, *Meno* 86d–87c.

36. For more on Plato's invention of the term *rhêtorikê* to characterize and attack the Sophists, see Schiappa, "Did Plato Coin Rhêtorikê?"

37. Latour's *Pandora's Hope* is a notable exception, and I engage with his work in some detail in the next chapter.

38. Ibid., 248.

39. Balaguer, "Realism and Anti-realism"; see also Balaguer, *Platonism and Anti-Platonism*.

40. Burton, "Practices of Mathematicians," 140.

41. See Boaler, "Mathematics from Another World"; Rav, "Critique"; and Ho and Hedberg, "Teachers' Pedagogies."

42. See Hanan, Ghosh, and Brook, "Banking on the Present," 141; and Reyes, "Algorithms and Rhetorical Inquiry."

Chapter 2

1. B. Russell, "Study of Mathematics."
2. B. Russell, *Power*, 314.
3. Davis and Hersh, *Descartes' Dream*, 57.
4. Perelman's work showed the Platonic roots of modernist thought; see especially *Realm of Rhetoric*. Latour's work (see especially *We Have Never Been Modern*) is strong in terms of modernism's impact on understandings of science.

5. Rotman, *Ad Infinitum*, 5. See also Rotman, *Signifying Nothing*; and Rotman, *Mathematics as Sign*.

6. Rotman, *Mathematics as Sign*, 30.

7. On political realism, see Wright, *Realism, Meaning, and Truth*; Hariman, *Political Style*; and Reyes, "Swift Boat Veterans."

8. Rotman, *Mathematics as Sign*, 29.

9. Frege, "Thought," 307; see also Frege, "On Sense and Reference."

10. I draw mostly on Rotman's more recent *Mathematics as Sign* because there one finds the clearest articulation of his approach.

11. Rotman, *Mathematics as Sign*, 13.

12. Ibid., 14.

13. Ibid.

14. Ibid., 17.

15. Ibid., 18.

16. The issue of the relationship between informal and formal mathematical discourse, the discursive and argumentative strategies within each, and the rhetorical purposes of each remains an unexplored and potentially rich area for rhetorical analysis. The best treatment of argument and informal mathematics I'm aware of is still Lakatos, *Proofs and Refutations*.

17. For others who make this argument, see Thurston, "On Proof and Progress"; Lakoff and Núñez, *Where Mathematics Comes From*; and Lakatos, *Proofs and Refutations*.

18. Rotman, *Mathematics as Sign*, 35.

19. Ibid., 121.

20. Regarding computers and mathematics, an interesting phenomenon has emerged in the twenty-first century: powerful computers are analyzing enormous data sets and are producing complex mathematical formulas out of those data sets that even the best mathematicians cannot *understand*—they know they work to predict certain phenomena in the data set but they cannot give meaning to those predictions. The fact that computers can generate novel mathematical formulas significantly undermines the Platonic view of mathematics, for there is no recollecting soul to be dialectically mined (see Rotman, *Mathematics as Sign*, 126–28).

21. Lakatos's work reveals the importance of historical context and the dynamics of argumentation in mathematical innovation. See *Proofs and Refutations*.

22. Wigner, "Unreasonable Effectiveness."

23. Rotman, *Ad Infinitum*, 140, 141.

24. On mathematics and capitalism, see chapter 5 of this book as well as Hanan, Ghosh, and Brook, "Banking on the Present"; Chaput and Colombini, "Mathematization of the Invisible Hand"; and I. Ascher, *Portfolio Society*. For insightful accounts of the emergence of Greek geometry and its debt to empirical, material features of the world, see Serres, "Gnomon"; and Netz, *Shaping of Deduction*.

25. Lakoff and Núñez, *Where Mathematics Comes From*, xi.

26. Ibid., 5.

27. Ibid., 2–3.

28. Ibid., 6.

29. Ibid., 385.

30. To the skeptical reader who thinks math is only metaphorical at the basic level: nearly half of Lakoff and Núñez's *Where Mathematics Comes From* addresses more complex mathematics, offering analyses of the concept of infinity and of Euler's classic equation: epi + 1 = 0. We explore the invention of imaginary numbers and the significance of this equation in chapter 4.

31. See Schiralli and Sinclair, "Constructive Response," 84.

32. Plutarch, *Marcellus' Life* 14.7, 14.8, 14.9.

33. Recall Russell's claim: "Mathematics takes us still further from what is human, into the region of absolute necessity, to which not only the actual world, but every possible world, must conform" ("Study of Mathematics"). Russell's mathematical realism, which is the dominant philosophy of math since Plato, tells of a world of abstract but unchanging objects hidden from quotidian view. For further evidence of realism's continued influence within mathematics, see Putnam's famous *Mathematics, Matter, and Method* as well as Resnik's widely regarded *Mathematics as a Science*.

34. Nussbaum shows how geometry emerged for thinkers like Parmenides and Plato as an alternative to *doxa* and *tuchē* (opinion and luck/chance); see *Fragility of Goodness*, 110.

35. Latour, *We Have Never Been Modern*, 110.

36. For economy's sake I dare not go too far into the woods on this point. Suffice it to say that whole books have been written teasing out the implications of mathematical realism on everything from mathematics itself to mathematics pedagogy to the ontological positioning of the human vis-à-vis nature to the relationship between democracy and the mathematical and natural sciences. For those interested in pursuing such topics from a rhetorically friendly perspective, see Rotman, *Ad Infinitum*; Rotman, *Signifying Nothing*; Rotman, *Mathematics as Sign*; MacKenzie, *Engine, Not a Camera*; Latour, *Science in Action*; Latour, *Pasteurization of France*; Latour, *Pandora's Hope*; and Barad, *Meeting the Universe Halfway*.

37. Latour, *We Have Never Been Modern*, 110.

38. Barad, *Meeting the Universe Halfway*, 71–94.

39. Latour's position here is consistent with his broader philosophical rejection of modernist metaphysics, which emerged historically with the rise of Cartesianism, modern algebra, and a renewed commitment to mathematical realism (see Cifoletti, "Mathematics and Rhetoric"; Latour, *We Have Never Been Modern*; and Latour, *Pasteurization of France*.

40. Latour, *We Have Never Been Modern*, 129.

41. Latour, *Science in Action*, 234, 244.

42. Ibid., 223, 242–47.

43. Latour, *We Have Never Been Modern*, 137.

44. On these topics a few touchstones are Gould, *Mismeasure of Man*; Porter, *Trust in Numbers*; Mudry, *Measured Meals*; Jack, *Science on the Home Front*; O'Neil, *Weapons of Math Destruction*; and Finn, *What Algorithms Want*. See Wynn and Reyes, *Arguing with Numbers*, for a more comprehensive bibliography.

45. Barad, *Meeting the Universe Halfway*, 184.

46. Reyes, "Rhetoric in Mathematics," 166–67.

47. The notion of rhetoric as a multiplicity (rhetorics) rather than a singularity as well as entailing the study of not just symbolic action but the entanglements of symbolicity and materiality that compose our worlds is very much in line with new materialist theories of rhetoric—especially Stormer's notion of rhetoric as polythetic (see "Rhetoric's Diverse Materiality" and "Articulation"). Stormer's idea of rhetoric as polythetic is deeply indebted to Barad's *Meeting the Universe Halfway*. A fuller explication of a constitutive approach to mathematical discourse can be found in Rotman's work (see especially *Ad Infinitum* and *Signifying Nothing*) as well as in Mudry, *Measured Meals*; and Reyes, "Rhetoric in Mathematics." Charland's work offers an elaboration of the theory of constitutive rhetoric (see "Constitutive Rhetoric"). Barad's *Meeting the Universe Halfway* shows how a constitutive approach challenges several intellectual orthodoxies, including the central tenet of representationalism that assumes words and things are both independent and determinate; the atomistic metaphysical belief that the world is composed of individual entities with definite boundaries and characteristics; and the foundationalist faith in the separability of knower and known concomitant with the

notion that proper experimental measurement reveals the intrinsic properties of independently existing objects (107).

48. I am not advancing an argument for symbolic or technological determinism here but rather a balanced approach that treats human, discursive, and nonhuman agencies symmetrically. The work of Latour (*Pandora's Hope*; *We Have Never Been Modern*), Haraway (*Simians, Cyborgs, and Women*; *Primate Visions*), Condit ("Race and Genetics"), and Barad (*Meeting the Universe Halfway*) are exemplars of the kind of symmetrical analysis I have in mind.

49. See the special issue of *Philosophy and Rhetoric*: Davis and Ballif, "Pushing the Limits" as well as the editors' introduction: Davis and Ballif, "Introduction."

50. See Lakatos, *Proofs and Refutations*; Hacking, *Philosophy of Mathematics*; and Cellucci, "Philosophy of Mathematics."

Chapter 3

This chapter is derived in part from Reyes, "Rhetoric in Mathematics," © Taylor and Francis.

1. Throughout the book I capitalize Newton's and Leibniz's methods to distinguish them from calculus in the generic sense. I also refer to "Newton's and Leibniz's Calculus," but this is meant to simplify presentation and not to indicate that Newton and Leibniz worked collaboratively on developing the Calculus. Quite to the contrary, they were bitter rivals (see Hall, *Philosophers at War*). This is not of importance here because this work focuses on the constitutive rhetoric that animates and gives "substance" to the concept of the infinitesimal and how that "substance" calls forth the situational rhetoric that Newton and Leibniz engaged. In any case Newton and Leibniz developed their own distinctive terminologies out of significantly different contexts (Newton in Britain and Leibniz on the continent), but these differences in terminology cannot be linked directly to a contest over origination. See Newton, *Method of Fluxions*; and Leibniz, "New Method."

2. Leibniz's approach was less geometrical and more algebraic in formulation. Thus, to obtain the derivative dy from $y = x^2$, Leibniz used e for dx to get $(x + e)^2 - x^2$ or $2xe + e^2$ and then just dropped e^2 (the square of an infinitesimal error); therefore, $dy = 2xe$.

3. Prior to the Calculus, Euclidean geometry provided the method of exhaustion, but this method was both cumbersome and limited to only the simplest of geometric objects. The Calculus, on the other hand, could approximate any variety of complex curvatures, which made it ideal for the study of rates of change.

4. Eves, *Great Moments*, 23.

5. I am offering a description of rhetorical practice and not rhetoric as such.

6. Heidegger, "Modern Science," 262.

7. Desilet, "Physics and Language," 343.

8. I do not intend to reinvigorate a substance/appearance dichotomy; I mean to explode that binary by exploring a mathematical concept that is *excessive* to such thinking. Gaonkar's "Idea of Rhetoric" claims that rhetoric often is positioned parasitically by scholarship in the rhetoric of science. Here I show that rhetoric is the "material" out of which a whole new system of mathematics emerges.

9. I am not claiming that Newton and Leibniz invented mathematical rhetoric, but that the use of infinitesimals called forth several forms of rhetoric.

10. Davis and Hersh, "Rhetoric and Mathematics," 54.

11. See Merriam, "Words and Numbers"; and Ernest, "Forms of Knowledge," respectively.

12. Rotman's work on semiotics and mathematics is an exception to this rule. Rotman attempts to disclose the semiotic ungroundedness of certain foundational mathematical concepts. See Rotman, *Ad Infinitum*; and Rotman, *Signifying Nothing*.

13. The unique rhetorical status of the infinitesimal begs the question of the rhetorical status of mathematics in general. Although that topic is beyond the scope of this project, I would agree (at least initially) with Rotman (*Ad Infinitum; Signifying Nothing*) that mathematics operates on an ideal plane and is informed by a Platonic philosophy that has its own rhetorical features. That said, most mathematical concepts emerge from within the boundaries of the established mathematical logic of the time (that is, from the rules that govern the action allowable in the mathematical world). Once in a while, however, a mathematical concept comes along that is so unique that it transgresses the boundaries of mathematical practice, often requiring novel rhetorical arguments as its substance and its advocate.

14. Euclid, *Thirteen Books*, 114.

15. For an extended treatment of these issues, see Jesseph, "Philosophical Theory."

16. Burke, *Rhetoric of Motives*. This is intended to describe how the concept of the infinitesimal (its rhetorical makeup) called forth the situational rhetoric explored in the following pages. In a certain sense both layers of rhetoric are "constitutive" of the infinitesimal, but I am using *constitutive* here to designate the rhetoric that governs the mathematical meaning of the infinitesimal, bringing up the debates over the Calculus during the seventeenth and eighteenth centuries.

17. On the concept of epistemes as they relate to discourses of power and discursive formations, see Foucault, *Order of Things*; and Foucault, *Archaeology of Knowledge*.

18. See Berkeley, *Discourse Addressed to an Infidel*; and Nieuwentijdt, cited in Boyer, *History of the Calculus*, 213.

19. Given the arguments in the previous chapter, this chapter may appear somewhat traditional in its analytic approach. The book, however, is structured such that as one moves deeper into mathematical genealogy and practice, one also moves farther and farther from conventional forms of rhetorical analysis—from, for instance, more conventional forms of analysis in the study of debates between Newton, Leibniz, and their contemporaries to a more unconventional analysis of the discursive fabrication of mathematical relations and from the analysis of human agents and agency to an increasing focus on nonhuman agents and agency. At the same time, we also progressively move from the technical spheres of mathematical discourse to an increasing emphasis on the translative force of mathematical discourse in the social-material world, which demands a symmetrical treatment of the technical practices of mathematical production and the translative modes of implementation within a social collective.

20. Berkeley leveled a scathing critique precisely along these lines, arguing that if one asks what "things" infinitesimals express, one "shall discover much Emptiness, Darkness, and Confusion; nay, if I mistake not," Berkeley goes on, infinitesimals create "direct Impossibilities and Contradictions" (*Analyst*, 4).

21. Rotman, *Ad Infinitum*, 41–42.

22. Bishop, "Semantic Flexibility," 225.

23. On the clarity of the Calculus, Bernard de Fontenelle, a highly respected French mathematician and philosopher, wrote, "It [the Calculus] does not cease still to cast us into the abyss of a profound darkness, or at the very least into realms where the daylight is extremely weak" (*Éléments de la géométrie*, iv).

24. Havelock speaks of a similar concept when he discusses the "stretching" of language toward the abstract by the pre-Socratics. See "Linguistic Task."

25. Rotman, *Ad Infinitum*, 145, 44.

26. See Nieuwentijdt, who began the criticisms of the Calculus, in Boyer, *History of the Calculus*, 213; Fontenelle, *Éléments de la géométrie*; and Raphson, *History of Fluxions*.

27. Along these lines, Berkeley writes, "For to consider the proportion or *Ratio* of things implies that such things have Magnitude"; but to do this with infinitesimals, he argues, "is to

talk unintelligibly" (*Analyst*, 15). *The Analyst* is considered by many the most powerful critique of the Calculus; see also Luce, *Berkeley and Malebranche*.

28. Boyer, *History of the Calculus*, 209.

29. Newton's treatment of infinitesimals in the *Principia* changed throughout his career, but he did not enjoy spirited philosophical exchange or criticisms of his work, especially those he felt were based on peripheral issues.

30. Boyer, *History of the Calculus*, 218.

31. Newton has been quoted by many as claiming later in his career that "the very smallest errors in mathematical matters are not to be neglected" (Cajori, appendix to Newton, *Sir Isaac Newton's Mathematical Principles*, 654).

32. Newton, cited in Burtt, *Metaphysical Foundations*, 223.

33. There are two forms of foundationalism, one scientific and concerned primarily with strict empiricism and the other mathematical and concerned primarily with strict adherence to Euclidean geometry. Cristoph Clavius, an important astronomer and mathematician of the early seventeenth century, expresses well the perspective of mathematical foundationalism: "The theorems of Euclid and the rest of the mathematicians, still today as for many years past, retain in the schools their true purity, their real certitude, and their strong and firm demonstration... and thus so much do the mathematical disciplines desire, esteem, and foster truth, that they reject not only whatever is false, but even anything mere probable, and they admit nothing that does not lend support and corroboration to the most certain demonstrations" (Lattis, *Between Copernicus and Galileo*, 35). This was not a univocal position of all seventeenth-century mathematicians (Cavalieri's *Geometria indivisibilibus continuorum* figures as an early, although tentative, departure from mathematical foundationalism); it was, however, the generally accepted view of rigor in mathematics that caused distrust of the Calculus among scientists and mathematicians. As Jesseph remarks, "They [infinitesimals]... violated widely accepted philosophical canons which declared the infinite to be incomprehensible and infinitesimal methods inadmissible" ("Philosophical Theory," 215). In addition, Kline argues, "Many of the English mathematicians, perhaps because they were in the main still tied to the rigor of Greek geometry, distrusted all the work on the calculus" (*Mathematical Thought*, 389).

34. Cajori, appendix to Newton, *Sir Isaac Newton's Mathematical Principles*, 654.

35. De Morgan, "On the Early History," cited in Cajori, appendix to Newton, *Sir Isaac Newton's Mathematical Principles*, 653, 654.

36. Cajori, appendix to Newton, *Sir Isaac Newton's Mathematical Principles*, 654.

37. Ibid.

38. Boyer, *History of the Calculus*, 213.

39. Newton, *Isaaci Newtoni opera*, 250–51, 40–41.

40. Newton, *Sir Isaac Newton's Mathematical Principles*, 38.

41. Kline suggests that "neither Newton or Leibniz clearly understood nor rigorously defined his fundamental concepts" (*Mathematical Thought*, 384). Stewart argues that Newton's "logic seems faulty. Over an interval of zero seconds, the position changes by zero, so the calculation becomes 0/0, and every mathematician knows that 0/0 can be anything you like" ("Sweet Nothings," 27). A somewhat commonplace conclusion among contemporary scholars is that Newton and Leibniz found the concept of the infinitesimal extremely useful but did not clearly understand it.

42. Criticism ranged from Fontenelle's rather mild comments to Berkeley's indictment of infinitesimals as absurdities. Berkeley asked, "And what are these Fluxions? The Velocities of evanescent Increments? And what are these same evanescent Increments? They are neither finite

Quantities nor Quantities infinitely small, nor yet nothing. May we not call them the Ghosts of departed Quantities?" (*Analyst*, 18).

43. Newton, *Sir Isaac Newton's Mathematical Principles*, 38.

44. Berkeley supports the reading that Newton's textual changes were a response to the problems caused by the infinitesimal: "It is curious to observe, what subtlety and skill this great Genius [Newton] employs to struggle with an insuperable Difficulty; and through what Labyrinths he endeavors to escape the Doctrine of Infinitesimals" (*Analyst*, 8).

45. See Boyer, *History of the Calculus*, 210–15.

46. Newton, *Isaac Newton's Papers and Letters*, 314–15.

47. One can demonstrate geometrically the decrease in error as the number of rectangles estimating the area under a curve increases (as with figure 5), but what one cannot do geometrically is take this process to its logical and intuitive conclusion—that is, sum up infinitesimal rectangles. As a result, one cannot prove the Calculus with mathematical rigor as long as one's notion of rigor is based on Euclidean geometry.

48. Newton, *Sir Isaac Newton's Mathematical Principles*, 397.

49. Boyer, *History of the Calculus*, 219.

50. Ibid., 209.

51. Leibniz, *Philosophische schriften*, 4:218.

52. Leibniz, *Leibniz Selections*, 185.

53. Leibniz, *Early Mathematical Manuscripts*, 147.

54. Boyer, *History of the Calculus*, 213.

55. Leibniz, *Philosophische schriften*, 6:90.

56. Berkeley's *Analyst* offers the most substantial critique of this; see also Jesseph, "Philosophical Theory."

57. Child, "Critical and Historical Notes," 149–50.

58. Leibniz, *Early Mathematical Manuscripts*, 155.

59. Berkeley, *Analyst*, 16.

60. Child, "Critical and Historical Notes," 147.

61. Leibniz, *Leibniz Selections*, 188, 71.

62. Benardete, *Infinity*, 18.

63. Leibniz, *Opera Omnia*, 500.

64. Leibniz, *Leibniz Selections*, 72.

65. Vilenkin, *In Search of Infinity*, 12–13.

66. Scientists and mathematicians of the seventeenth century generally accepted Euclidean forms of mathematical proof as sufficient verification because each symbol in Euclidean geometry has a corresponding geometric object to which it refers; that abstract reference to a spatial relation can then easily be mapped onto physical processes and verified empirically. Infinitesimals presented a special problem because they could not be represented spatially.

67. Newton's work testifies to this, as he first completed the *Principia* using his method of fluxions and then translated his work back into acceptable forms of Euclidean geometry. Cavalieri makes a similar effort (although to a lesser degree) in the second half of his *Geometria degli indivisibili*.

68. Boyer, *History of the Calculus*, 227. This is an incredible shift in perspective—a retreat from or rejection of the physical world as the primary "criterion of truth" in mathematics as well as the establishment of a hierarchy of truth on which "inner consistency" in mathematics enjoyed the highest regard. From this perspective mathematical truth is more than "natural law"; it is excessive to it. Indeed, mathematical truth "deals with relations rather than physical

existence," including the relations that constitute what we call natural law. And, as math deals with those relations, it can sometimes, rarely but on occasion, reconfigure those relations or introduce a novel relation, dramatically transforming the social-material world in the process. This shift also suggests some precedence for the relational-network approach to math developed in this book within the mathematical community.

69. Stewart, "Sweet Nothings," 28.
70. See Bacon, *Philosophical Works*.
71. In *Mathematical Experience* Davis and Hersh quote Soloman Feferman, who attests to the decline of foundationalism in mathematics, admitting that "the mathematician at work relies on surprisingly vague intuitions and proceeds by fumbling fits and starts with all too frequent reversals" (357).
72. Rotman, *Ad Infinitum*, 157.
73. Newton and Cohen, *Isaac Newton's Papers*, 274.
74. Vilenkin, *In Search of Infinity*, 13.
75. For more on Bentley, see Newton and Cohen, *Isaac Newton's Papers*, 274.
76. See Foucault, *Order of Things*.
77. Vilenkin, *In Search of Infinity*, 13.
78. Heidegger, "Modern Science," 269.
79. Davis and Hersh, "Rhetoric and Mathematics."
80. Robinson, *Nonstandard Analysis*.

81. The Boltzmann equation, which describes how a cloud of tiny particles changes in density as the particles interact and helps to predict the motion of stars in the galaxy, was considered questionable by scientists because no one knew if it would suddenly go haywire. No one could prove it was a stable equation. In 1984 Leif Arkeryd used nonstandard analysis to prove the stability of the Boltzmann equation. In addition, nonstandard analysis, which is based on infinitesimals, has helped scientists and mathematicians understand Brownian motion and the Jordan theorem and improve computer graphics.

82. Foundationalism is not "dead" among scientists and mathematicians, but it is no longer the only dominant view.

83. Rorty, *Contingency, Irony, Solidarity*, 6.

84. What haunts mathematics, in this sense, is not error—whether from the use of infinitesimals or some other mathematical concept—but instead the ways mathematical discourse can unleash the radically new and accelerate the world's differential becoming.

Chapter 4

1. Gantumur, "Complex Numbers."
2. Singh, "Imaginary Number."
3. Nahin, *Imaginary Tale*, 5–6.
4. Diophantus's *Arithmetica* is included in Heath, *Diaphantus of Alexandria*.
5. Netz, *Shaping of Deduction*.
6. On the Pythagoreans, see Philip, *Pythagoras and Early Pythagoreans*; and Burkert, *Lore and Science*.
7. See also Netz, *Shaping of Deduction*.
8. Ibid., especially chapter 2.
9. See Boyer and Merzbach, *History of Mathematics*, 282–89.
10. See Waerden, *History of Algebra*.
11. Bombelli, *Algebra*. See also Bombelli and Cataldi, "Bombelli and Cataldi."

NOTES TO PAGES 90–100   163

12. Boyer and Merzbach, *History of Mathematics*, 288.
13. Blank, "Imaginary Tale."
14. Boyer and Merzbach, *History of Mathematics*, 406.
15. Strogatz, *Joy of X*, 51.
16. Boyer and Merzbach, *History of Mathematics*, 442.
17. For those interested, Lakoff and Núñez take four chapters to thoroughly unpack this equality in *Where Mathematics Comes From*; see pages 383–451.
18. For those not familiar, trigonometry was originally the study of the relations between sides and angles of triangles and emerged in ancient Greece for purposes of astronomy. See Brummelen, *Mathematics of the Heavens*. Modern algebraic trigonometry, however, did not emerge in any robust form in Europe until the sixteenth and seventeenth centuries.
19. The idea of "concept-stretching" as central to the growth of mathematics comes from Lakatos's *Proofs and Refutations*, where he shows the crucial role it plays in the growth of mathematical knowledge. While I disagree with Lakatos's epistemic approach to mathematics, his continuity theory of its historical development, and the subtle forms of realism lurking in his work, I find many of his observations about mathematical practice insightful.
20. Lakoff and Núñez, *Where Mathematics Comes From*, 397.
21. Wallis's *Treatise of Algebra* (1685) offered a proof of the physical interpretation of negative numbers, which until the end of the 1600s were still viewed with suspicion. Building on the visual logic of the Cartesian coordinate system, he argued, in short, that a positive number simply means a distance measured from a zero point to the right, and negative numbers means a distance measured from a zero point to the left. In this argument, however, one can see the desire for demonstrative physical proof of mathematical concepts and thus the continued hegemony of the episteme of representation. See Wallis, *Treatise of Algebra*.
22. For a more detailed background on "e" (which is tangential to the focus of this chapter), see Lakoff and Núñez, *Where Mathematics Comes From*, 399–419.
23. Euler, cited in Boyer and Merzbach, *History of Mathematics*, 449.
24. Even Nahin's recent history of imaginary numbers titles the section on Euler "Wizard Mathematics" (see Nahin, *Imaginary Tale*, chapter 6). Gauss originally proved the fundamental theorem of algebra (with a small gap) in his 1799 dissertation.
25. See Gauss's *Disquisitiones arithmeticae*, in Gauss, *Werke*.
26. Wessel, *Analytical Representation of Direction*; Argand, *Imaginary Quantities*.
27. Boyer and Merzbach, *History of Mathematics*, 507.
28. To understand the mathematics of complex numbers here, we can put the original point $(3, 0)$ into the complex form $a + bi$, where a is the real component and bi is the imaginary component of complex numbers. Thus $(3, 0) = 3 + 0i$. Multiplying by i yields $i(3 + 0i)$, which equals $3i$ or, in complex form, $0 + 3i = (0, 3i)$.
29. While going through my mathematics courses in college, I had little sense that virtually all the math we learned after calculus was indebted in one way or another to imaginary numbers. Only now, in writing this book, do I see their vast significance both in mathematics and virtually every other technical field of modern society.
30. On Descartes's mechanical representationalism, see Rouse, *Engaging Science*, 209.
31. Barad, "Posthumanist Performativity," 811.
32. By "natural" I mean those phenomena (each of which is a network of relations) that existed long before the species *Homo sapiens*.
33. It is true that we might not absolutely need imaginaries to understand electricity and electrical engineering mathematically; however, imaginary numbers make the calculations infinitely more straightforward, such that humans could reasonably calculate them—a crucial element in a world without powerful computers. As Caroline Series (math professor, University

of Warwick) notes, "If . . . you didn't have the power of these imaginary numbers to do the calculations, it would take so impossibly long that even with modern computers you couldn't do it" (Sautoy and Series, "In Our Time," 39:01–20).

34. Ibid., 40:22–41:25.

35. See Haraway, *Simians, Cyborgs, and Women*; Haraway, "Promises of Monsters"; and Haraway, *Modest_Witness*.

Chapter 5

Segments of this chapter originally appeared in Reyes, "Algorithms and Rhetorical Inquiry," published by Michigan State University Press; and Reyes, "Horizons of Judgement."

1. While the *meaning* of algorithms is polysemous within contemporary discourse, one can define algorithms as simply an assemblage of rules used to automate the treatment of information. If A is true, then do B; if not, then do C. This is the basic "if/then/else" logic of algorithmic computing (see Ullman, "Programming the Post-human").

2. On issues of algorithmic inaccessibility, see Finn, *What Algorithms Want*, especially chapters 1 and 3.

3. Ibid., 7–13.

4. I develop the concept of the horizon of judgment in the body of this chapter. In short, it refers to the network of human and nonhuman assumptions, constraints, and delimitations necessary for a mathematical algorithm to have specificity and validity. I use the metaphor of horizon to underscore the distant but no less present domain of validity that constrains every mathematical algorithm. The notion that mathematical discourse is an active agent in the formation of economic realities builds on Hanan, Ghosh, and Brook's insight that mathematical modeling in economics "does not so much represent a marketplace that exists *a priori* as it simulates or 'performs' a world that is imagined in its models" ("Banking on the Present," 141).

5. This copula first emerged in Li, "On Default Correlation." The purpose of this chapter is to render economic algorithms more accessible and in the process enliven a critical discussion of them as active discursive agents in the composition of economic fields of power. As such, while this chapter also draws on insights from rhetoric of economics research, the primary focus of the analysis is on mathematical discourse within economics and the algorithms that emerge therein. Past research on economic rhetoric focuses primarily on traditional rhetorical practices within economics—that is, on speeches by prominent actors (such as Alan Greenspan) or on argumentation within economics and among economists (policy debates, for instance) or on the circulation of certain influential ideographs as carriers of conflicting ideologies. As with all tendencies, one can find exceptions to the rule (McCloskey's *Rhetoric of Economics* but also Hanan, Ghosh, and Brook, "Banking on the Present"), yet the preponderance of research in rhetorical studies has examined conventional rhetorical forms of economic discourse to the exclusion of the mathematical discourse that increasingly forms the bedrock of economic argument and practice. For an excellent entry into and expansive bibliography of the rhetoric of economics literature, see Hanan and Hayward, *Communication and the Economy*. Other touchstones include McCloskey, *Rhetoric of Economics*; Mirowski, "Shall I Compare"; Mirowski, *More Heat Than Light*; Aune, *Selling the Free Market*; Houck, *Rhetoric as Currency*; and Goodnight and Green, "Rhetoric, Risk, and Markets."

6. Human judgment has long been associated with the ability to take in particular information from one's context and make decisions based on a combination of that information and one's prior experience. This is the classical notion of human judgment one finds elaborated in Aristotle's notion of *phronesis*; see "Rhetoric" and "On Poetics"; see also Miller, "Presumption

of Expertise." Algorithmic automation displaces *phronesis* with abstract equations presumed to be free of subjective human judgment. As I elaborate later in the analysis, however, algorithmic automation is neither free of human judgment nor merely a means of spreading one particular set of human judgments. Instead, it is an amalgam of both human and nonhuman agencies.

7. Thomas, "Social Computing," 10.
8. Stiegler, *Technics and Time*, 80–81.
9. See Russell, Dewey, and Tegmark, "Research Priorities"; and S. Russell, *Human-Compatible Artificial Intelligence*.
10. Helbing et al., "Will Democracy Survive."
11. S. Russell, cited in Reid, "AI Boom," 16:24–32.
12. For more on these phenomena, see Zuboff, *Age of Surveillance Capitalism*.
13. Pasquale, *Black Box Society*, 9. For a recent review of critical algorithm studies, see Mittelstadt et al., "Ethics of Algorithms."
14. Pasquale and many other scholars see in the rise of an algorithmic culture the simultaneous and correlated rise of societies of control. On the structure and characteristics of societies of control, see Deleuze, "Postscript"; Hardt and Negri, *Empire*; and I. Ascher, *Portfolio Society*. On the recursive structure of complex "learning" algorithms, see Gillespie, "Algorithm"; and Neyland, "On Organizing Algorithms."
15. For an excellent collection of relevant literature on these issues, see the Social Media Collective reading list in "Critical Algorithm Studies."
16. O'Neil, cited in Upchurch, "To Work for Society." O'Neil details the rise of this opaque bureaucracy in her book *Weapons of Math Destruction*.
17. Pasquale, *Black Box Society*, 4.
18. See Mittelstadt et al., "Ethics of Algorithms," 12–14.
19. Lenglet espouses a widely shared view that "behind every algorithm lies a human," and while I agree that we need to be wary of the "romantic view" that grants algorithms "a life of their own," we also must address the nonhuman, nonindividualistic agencies constitutive of modern algorithms ("Algorithms and the Manufacture," 318).
20. For more on these rapidly transforming learning algorithms, see O'Neil, *Weapons of Math Destruction*; Johnson et al., "Abrupt Rise"; and Smith, "Franken-Algorithms."
21. Lenglet, "Algorithms and the Manufacture," 319. Even in Finn's book *What Algorithms Want*, which has many strengths, one finds nary an equation.
22. Overemphasis on algorithmic implementation also tends to treat algorithms as finished products, thus reinforcing the perception of them as cold, objective, and refined machine technologies instead of, as Gillespie notes, "in fact a fragile accomplishment" ("Relevance of Algorithms," 169). When scholars speak of algorithms, Gillespie observes, they often reify into a singularity what is in actuality a "complex sociotechnical assemblage" ("Algorithm," 24). A few scholars are beginning to address the assemblage of mathematical discourse constitutive of complex algorithms. See especially Kockelman, "Anthropology of an Equation"; and Thomas, "Social Computing."
23. For contemporary research along these lines, see Reyes, "Stranger Relations"; and Wynn and Reyes, *Arguing with Numbers*.
24. There are a few scholars that buck this trend, challenging the human-centric approach to algorithms (see Gillespie, "Relevance of Algorithms"; and Kockelman, "Anthropology of an Equation"), yet those challenges are few and far between and seemingly constrained by the absence of a critical method that effectively engages with the mathematical discourses constitutive of complex algorithms.
25. The risk in focusing on the construction of algorithms is that we, as critical scholars, lose sight of the massive cultural changes algorithms enable and effect. The hope of this book

is to create a balanced analytic that attends to both practices of assemblage and sociomaterial consequence.

26. Salmon, "Formula."

27. Jones, "Formula That Felled."

28. Greenspan, October 12, 2005, cited in Krugman, *End This Depression*, 54.

29. Salmon, "Formula," 19.

30. The discursive shift from "model" and "forecasting" to "technology" and "measurement" underwrites a shift from a fallibilist to a realist epistemology.

31. For more on the impact of MBSs on the rise of neoliberalism and the transformation of the home into an "abstract financial equation," see MacKenzie, *Engine, Not a Camera*; and Hanan, "Home Is Where the Capital Is."

32. The risk exposure changes because investors in MBSs are paid out hierarchically. For example, AAA investors would be at risk of losing money only if more than 20 percent of a mortgage pool defaulted, while that percentage would decrease with each tranche subordinate to the AAA tranche.

33. Hanan, Ghosh, and Brook's "Banking on the Present" offers an excellent overview of the rise of MBSs within a broader critique of neoclassical economics. For more on mortgage-backed securities and their relationship to subprime mortgages and the 2007 economic collapse, see Coval, Jurek, and Stafford, "Economics of Structured Finance"; Fabozzi, Bhattacharya, and Berliner, *Mortgage-Backed Securities*; and McLean and Nocera, *All the Devils*.

34. The basic mathematics behind default correlation are as follows: correlation is considered a probability function of the relationship between two debt obligations, meaning it ranges from 1 to −1, where 0.2 default correlation would represent a 20 percent probability that if one mortgage defaults the other will as well. In contrast, a −0.2 default correlation would represent a negative 20 percent probability that if one mortgage defaults the other will default, meaning that if one defaults the chances of the other defaulting as well decreases. For more background on problems of default correlation, see Coval, Jurek, and Stafford, "Economics of Structured Finance."

35. Coval, Jurek, and Stafford, "Economics of Structured Finance," 16.

36. As Coval, Jurek, and Stafford, note, the use of the Li copula to "repackage risks and to create 'safe' assets from otherwise risky collateral led to a dramatic expansion in the issuance of structured securities, most of which were viewed by investors to be virtually risk-free and certified as such by the rating agencies" ("Economics of Structured Finance," 3). Salmon elaborates, "Using Li's copula approach meant that ratings agencies like Moody's—or anybody wanting to model the risk of a tranche—no longer needed to puzzle over the underlying securities. All they needed was that correlation number, and out would come a rating telling them how safe or risky the tranche was" ("Formula," 19).

37. Li, "On Default Correlation," 2.

38. A basic tenet of realism is that to have unmediated access to "the real," one must cast off all ideology and artifice. Economic realists see the language of mathematics as the best means for casting off artifice, exposing ideology, and discovering the real (see Maki, "How to Combine"; and McCloskey, "Two Replies." Thus the discourse of realists purports to describe the world in plain language (sans artifice), positioning the author as witness to the world so described. Realism as a rhetorical style is well developed in Hariman's *Political Style* as well as in Reyes's "Swift Boat Veterans." On the pervasiveness of realism within economics, see Aune, *Selling the Free Market*. Li's article is highly conventional in this sense, providing ample evidence of a realist style throughout (see especially "On Default Correlation," 3–5).

39. Aune, *Selling the Free Market*, 40; see also Hanan, Ghosh, and Brook, "Banking on the Present."

40. Li, "On Default Correlation," 2.

41. While the mathematization of economics first began with the rise of probability and statistics in the 1940s and 1950s (see McCloskey, *Rhetoric of Economics*, 112), it was not until the 1970s that it became a significant force within structured finance, concomitant with the spread of computers and increasing computational power (see Finn, *What Algorithms Want*, especially chapter 5).

42. Li, "On Default Correlation," 2, 3.

43. For an extensive discussion of scalable versus nonscalable phenomena, see Taleb, *Black Swan*.

44. Federal Reserve Bank, "Delinquency Rate."

45. Some of these elements of his argument are explicit in Li's text (like the collective poor understanding of default correlation), and some are implicit (like the scalability or nonscalability of default correlation) but nevertheless necessary for the integrity of Li's mathematical treatment.

46. Note that for the purposes of this analysis we treat "function," "equation," and "algorithm" as synonymous even while acknowledging differences between these terms within the technical mathematical sphere.

47. Li, "On Default Correlation," 3–4.

48. Latour, *Science in Action*, 244.

49. See Porter, *Trust in Numbers*, 49–72; McLean and Nocera, *All the Devils*; and Krugman, *Return of Depression Economics*.

50. Li, "On Default Correlation," 6–8, 7.

51. Ibid., 16.

52. By "terraforming" I mean the ways math can transform our social-material worlds.

53. Finn, *What Algorithms Want*, 20.

54. On the parahippocampus, see Burgess, Maguire, and O'Keefe, "Human Hippocampus"; and Sparrow, Liu, and Wegner, "Google Effects on Memory"; on Netflix's algorithmic effects, see Amatriain, "Netflix Recommendations"; and Madrigal, "How Netflix Reverse-Engineered."

55. See Coval, Jurek, and Stafford, "Economics of Structured Finance"; and Lewis, *Big Short*.

56. The Li copula sponsored many other hybrids within the CDO market, such as cash, synthetic, and hybrid CDOs. For details on all the financial products that emerged in the wake of the Li copula, see McLean and Nocera, *All the Devils*.

57. Ibid., 123.

58. For a more in-depth explanation of pooling and tranching, see Coval, Jurek, and Stafford, "Economics of Structured Finance," 6–7.

59. Ibid., 8–9.

60. By way of example, take the junior tranche in our hypothetical loan pool and imagine combining it with other junior tranches from other CDOs or MBSs. Using the same logic, one could pool those debt obligations and tranche them, creating CDO^2 products with senior tranches that would be above investment grade. With each repooling and retranching process, however, the number of risky underlying debt obligations (like subprime mortgages) increases, yet the Li copula convinced many within finance that the particular characteristics of the underlying loans were irrelevant. As the Financial Crisis Inquiry Commission's *Financial Crisis Inquiry Report* later discovered, "Approximately 80% of these CDO tranches would be rated triple A despite the fact that they generally comprised the lower-rated tranches of mortgage-backed securities" (127).

61. Financial Crisis Inquiry Commission, *Financial Crisis Inquiry Report*, 130; McLean and Nocera, *All the Devils*, 201.

62. McLean and Nocera, *All the Devils*, 123; Benmelech and Dlugosz, "Credit Rating Crisis"; International Monetary Fund, "Historical GDP by Country."

63. Financial Crisis Inquiry Commission, *Financial Crisis Inquiry Report*, xvii.

64. Lewis, *Big Short*, 73.

65. Krugman, *Return of Depression Economics*, 155.

66. See Morgenson and Rosner, *Reckless Endangerment*, 283.

67. Lewis, *Big Short*, 73.

68. Coval, Jurek, and Stafford, "Economics of Structured Finance," 8–12.

69. Taleb, *Black Swan*, xxix. For more on Gaussian model fragility, see Haug and Taleb, "Option Traders Use."

70. I borrow the term *sliced* from Barad, who uses it to describe the power of new knowledge as the power to slice social-material relations into novel configurations (see *Meeting the Universe Halfway*, 183–85).

71. Finn, *What Algorithms Want*, 49.

72. There are many ways in which numbers can influence one's perception: the simplest example emerges from an experiment in which Amos Tversky and Daniel Kahneman—who first studied the anchoring effects of numbers—asked participants to guess "the percentage of African countries in the UN." One group spun a "wheel of fortune" rigged to stop at ten, then guessed the percentage. Another group spun the same wheel rigged to stop at sixty-five. The results were surprising: the first group's average guess was 25 percent and the second group's was 45 percent. Anchoring, Kahneman explains, "occurs when people consider a particular value for an unknown quantity before estimating that quantity" (*Thinking Fast and Slow*, 119).

73. On anchoring within disjunctive systems, see Tversky and Kahneman, "Judgment Under Uncertainty."

74. Kahneman, *Thinking Fast and Slow*, 428.

75. Anchoring is only one dimension of the copula's agential force, and it would not have had the impact it did in isolation. The copula's reduction of default correlation to a constant was buttressed by other dimensions of mathematical realism within mathematical discourse—by the widespread acceptance of the concept of equality within the realm of probability, for example; the deductive form of mathematical discourse, which begins with assumptions but ends with necessity; or the generalization of correlation, which is inherently descriptive of *particular* relations. While these points are beyond the scope of this book, they bear additional critical attention.

76. Stormer, "Rhetoric's Diverse Materiality," 306.

77. This approach is both pragmatic and philosophically disinclined to adoption of a single method, for the point of analysis here is not merely to reveal the hidden or to illuminate (though those are crucial elements of analysis) but, more important, to multiply our critical capacities for address and encounter with mathematical algorithms, whether within economics as a field of power or within other domains. On analysis as a force of multiplication, see Latour, *Reassembling the Social*, 144.

Chapter 6

1. See Isocrates, *On the Peace*.

2. For a latest compilation of scholarship on rhetoric and mathematics and a close analysis of Aristotle's influence in solidifying the original Platonic division, see Wynn and Reyes, *Arguing with Numbers*.

3. As Wynn and Reyes show in *Arguing with Numbers*, Aristotelian thought further developed and codified Plato's division between rhetoric and math. See Wynn and Reyes, "From Division to Multiplication."

4. See especially Little, "Rhetoric and Mathematics"; and Fahnestock, "New Mathematical Arts of Argument."

5. Nietzsche's critique of these human metaphysical conceits remains among the best and most incisive. See Nietzsche, *Philosophy and Truth*; see also Nietzsche, *Philosophy of Nietzsche*.

6. See Rav, "Critique"; and Ernest, *Social Constructivism*.

7. The point to make here is that mathematics is not abstract or rule driven by nature, but it can and often is *taught* as an abstract form of logical (rule-driven) reasoning. See Boaler, "Mathematics from Another World."

8. Ho and Hedberg, "Teachers' Pedagogies."

9. Barad, "Posthumanist Performativity," 804.

10. In Barad's terms, "Representationalism separates the world into the ontologically disjoint domains of words and things, leaving itself with the dilemma of their linkage such that knowledge is possible" ("Posthumanist Performativity," 811).

11. Note that when we use the term *mediation* here we are not using it in the traditional representationalist sense, in which inscriptions mediate between the world and the human mind in a way that renders that world intelligible. Rather, we use it here within a relational paradigm, in which inscriptions are media of translation and assemblage that can occasionally give rise to a novel principle of composition with the translative force necessary to transform the relations of the networks that compose our world.

12. Ceccarelli, "Rhetorical Criticism," 315–16.

13. As an exemplar of this discomfort and simultaneous acknowledgment of reality's plasticity, Depew and Lyne write of the importance of nature's recalcitrance as a check on the extremes of social constructivism, even as they recognize how increasing understanding of the networks of DNA have opened many previously recalcitrant dimensions of life to translation and transformation. See "Productivity of Scientific Rhetoric."

14. Walsh et al., "Forum," 403; Prior to this special issue, one might also look to Simonson's relatively early effort to address the influence and potential of new materialist theories for rhetoric; see "Review Essay."

15. The issue here, as Gross observed in a recent interview, is simply that "a technology, namely classical rhetoric, that was designed to teach young boys on how to give speeches in public forums, was not going to be a technology that could be easily transferred into a critical methodology, especially one of science, which was a whole lot more complex. And so what you had was a situation where you had to rethink and expand the notion of what kind of critical apparatus you would need" ("ARST Oral History Project," 10:45–13:42); see also Gross and Keith, *Rhetorical Hermeneutics*. This work to expand our theoretical and critical apparatus as rhetorical scholars is precisely what many new materialist theories of rhetoric are about and, more to the point, what the theory of translative rhetoric is about.

16. Graham, *Where's the Rhetoric*, 1.

17. By "micropractices" I mean to underscore all the practices of inscription involved in the articulation of mathematical statements with ontological force. Both chapters 4 and 5 trace these micropractices, the former in the context of the invention of imaginary numbers and the latter in the context of the 2008 financial crisis and the rise of algorithmic culture.

18. Kuhn, *Structure of Scientific Revolutions*.

19. Miller, "Novelty and Heresy," 501–2.

20. In Depew's and Lyne's words, "Just as we do not think creationists should colonize evolutionary biology by making it conform to their worldview, neither do we think that literary

humanists should be allowed to do something similar. We are afraid that social constructionism has fallen into that trap" ("Productivity of Scientific Rhetoric"). For similar concerns, see Condit, "Mind the Gaps."

21. Collier points out the absence of and need for a broader "philosophical vision to synthesize its [rhetoric of science and technology's] knowledge into a coherent story" ("Reclaiming Rhetoric of Science," 295). Perhaps the theory of translative rhetoric can offer such a "moment of synthesis" (297). Here the impulse would not be to merely prove rhetoric's presence in and importance to scientific and mathematical practice but instead to show how the translative force of scientific and mathematical discourse has increasingly *opened the world to rhetorical practice*—to the practices of writing and rewriting, coding and decoding, decomposing and recomposing. This process is at the heart of the "life is code" project at CERN and so many other twenty-first century scientific institutions (see Enriquez and Gullans, *Evolving Ourselves*).

22. For extensive treatments of both these points, see Fuller and Collier, *End of Knowledge*; Collier, "Reclaiming Rhetoric of Science"; and Graham, *Where's the Rhetoric*.

23. One task for rhetorical inquiry into math, in other words, is to not only understand how math grows and evolves but, equally important and derivative of that growth, understand the increasing ontological force of mathematical discourse, the likely paths that ontological force might take, and what social collectives might do to guide that ontological force.

24. Anthropocene, "Working Group on the Anthropocene." See also Zalasiewicz et al., "New World." Thinking translatively, when I say "rhetorical situation" what I mean is the network of relations that configure the social-material world.

25. The story of the increasingly complex entanglements between mathematics and capitalism is an incredibly important topic that is well beyond the scope of this book. Suffice it to say that much could and should be written about the ways mathematical apparatuses like the Cartesian coordinate system enabled and accelerated the grid-like commodification of geography and ecology. For more on this relationship from a rhetorical perspective, see Chaput and Colombini, "Mathematization of the Invisible Hand"; and Hanan, Ghosh, and Brook, "Banking on the Present."

26. On Burke's idea of rhetoric as a means of overcoming biological division, see Burke, *Rhetoric of Motives*, especially the sections on identification and consubstantiality.

27. Barad, "Posthumanist Performativity," 802. For an expansion of the ideas in this work, see Barad's *Meeting the Universe Halfway*.

28. "In part" is key here, for there are myriad other forces that have also contributed to the emergence of the Anthropocene, not least of which is the rise of industrial and late capitalism.

29. Some might argue that these phenomena would simply have emerged by different means, but I agree with mathematician Caroline Series when she argues that if we "didn't have the power of these imaginary numbers to do the calculations [for everything from electrical current to quantum motion] it would take so impossibly long that even with modern computers you couldn't do it" (Sautoy and Series, "In Our Time," 37:44–39:20). And, of course, we would not have modern computers either.

30. In fact, this revision of classical understandings of gravity are already well underway, as Srinivas Bettadpur, the director of the Center for Space Research, observes: "The classic idea of gravity being something that you measure once is no longer accepted. Gravity is an element that scientists must continue to monitor" because it is a network of relations continuously in evolutionary flux ("Matter in Motion").

31. Bennett, *Vibrant Matter*, 116–17.

32. Physics disproved the atomistic theory of the world long ago thanks, in part, to Neils Bohr's work. "Bohr's philosophy-physics," Barad reminds us, "poses a radical challenge not only to Newtonian physics but also to Cartesian epistemology and its representationalist triadic

structure of words, knowers, and things. Crucially, ... Bohr rejects the atomistic metaphysics that takes 'things' as ontologically basic entities. For Bohr, things do not have inherently determinate boundaries or properties, and words do not have inherently determinate meanings. Bohr also calls into question the related Cartesian belief in the inherent distinction between subject and object, and knower and known" ("Posthumanist Performativity," 813).

33. Fausto-Sterling, "Bare Bones of Race"; Fausto-Sterling, "Bare Bones of Sex"; Condit, "Race and Genetics."

34. For an excellent book about the ways the genetics revolution is merging with the nanotechnological revolution and the information revolution to potentially enable the fundamental transformation of biology and ecology, see Kurzweil, *Singularity Is Near*.

35. Other scholarly works warning against unification include Stormer, "Rhetoric's Diverse Materiality"; Stormer and McGreavy, "Thinking Ecologically"; Vivian, *Being Made Strange*; Davis, "Rhetoricity at the End"; Davis and Ballif, "Introduction"; Benoit-Barné, "Socio-technical Deliberation"; Greene, "Rhetoric and Capitalism"; Biesecker and Lucaites, *Rhetoric, Materiality, and Politics*; and Rickert, *Ambient Rhetoric*. As a catalyst for these conversations, see Gaonkar's original essay ("Idea of Rhetoric"), as well as the expanded version and various scholarly responses in Gross and Keith, *Rhetorical Hermeneutics*.

36. This is my understanding of Stormer's notion of "polythetic rhetoric" ("Rhetoric's Diverse Materiality").

37. This is true of most, if not all, symbolic action, as Lakatos's *Proofs and Refutations* shows with the history of the mathematical study of polyhedra, and Thurston has shown in "On Proof and Progress" with concepts like "derivative."

38. If one presumes (implicitly or explicitly) that rhetoric, as a unique entity or phenomenon, is "big"—that is, of broad influence—but fails to acknowledge its diversity, it becomes more difficult to see the specific translative rhetorical force of particular modalities of discourse, which then makes it more difficult to demonstrate the translative force of discourse itself (whether mathematical or otherwise).

39. See Dyson, "Biological and Cultural Evolution."

40. There is a story to be told here of the increase of influence of cultural evolution as it relates to the role mathematical discourse had in helping catalyze the transition from an agrarian to an industrial-capitalist culture that is tangential to the central arguments of this book. That said, one can surmise, along with many others, that collusion between mathematics and capitalism has been essential to the rise of biocultural evolution and the emergence of the Anthropocene. See O'Neil, *Weapons of Math Destruction*; Stockman, *Great Deformation*; and I. Ascher, *Portfolio Society*. Of course, race and colonialism cannot be ignored here: a few starting points on the history of collusion between race, colonialism, and capitalism might include Mbembé and Dubois, *Critique of Black Reason*; Mbembé and Corcoran, *Necropolitics*; and Reyes and Chirindo, "Theorizing Race."

41. See Kurzweil, *Singularity Is Near*.

42. Bennett, *Vibrant Matter*, 13. As Bennett astutely observes, "To harm one section of the web may very well be to harm oneself. Such an enlightened or expanded notion of self-interest is *good for humans*. . . . A vital materialism does not reject self-interest as a motivation for ethical behavior, though it does seek to cultivate a broader definition of self and of interest" (13).

Bibliography

Abbott, Darek. "The Reasonable Ineffectiveness of Mathematics." *Proceedings of the IEEE* 101 (2013): 2147–53.
Amatriain, Xavier. "Netflix Recommendations: Beyond the Five Stars." *Netflix Tech Blog.* April 6, 2012. http://techblog.netflix.com/2012/04/netflix-recommendations-beyond-5-stars.html.
Anthropocene. "Working Group on the Anthropocene." Subcommission on Quaternary Statigraphy. May 21, 2019. http://quaternary.stratigraphy.org/working-groups/anthropocene.
Argand, Jean-Robert. *Imaginary Quantities: Their Geometrical Interpretation.* Translated by A. S. Hardy. 1806. Reprint, New York: Van Nostrand, 1881.
Aristotle. *The Metaphysics.* Translated by Hugh Tredennick and George Cyril Armstrong. London: Heinemann; New York: Putnam's Sons, 1933.
———. *Posterior Analytics.* Vol. 1 of *The Complete Works of Aristotle: The Revised Oxford Translation.* Translated by Jonathan Barnes. Princeton: Princeton University Press, 1984.
———. *The Rhetoric.* Translated by W. Rhys Roberts. New York: McGraw-Hill, 1984.
———. *"Rhetoric" and "On Poetics."* Translated by W. Rhys Roberts. Edited by W. D. Ross. Franklin Center, PA: Franklin Library, 1981.
Arndt, Jörg, and Christoph Haenel. *Pi-Unleashed.* Berlin: Springer, 2001.
Ascher, Ivan. *Portfolio Society: On the Capitalist Mode of Prediction.* New York: Zone Books, 2016.
Ascher, Marcia. *Ethnomathematics: A Multicultural View of Mathematical Ideas.* Pacific Grove, CA: Brooks/Cole, 1991.
Asimov, Isaac. Foreword to Boyer and Merzbach, *History of Mathematics,* vii–viii.
Aune, James A. *Selling the Free Market: The Rhetoric of Economic Correctness.* New York: Guilford Press, 2001.
Bacon, Francis. *The Philosophical Works of Francis Bacon.* Edited by J. M. Robertson. Freeport, NY: Book for Libraries Press, 1970.
Balaguer, Mark. *Platonism and Anti-Platonism in Mathematics.* Oxford: Oxford University Press, 1998.
———. "Realism and Anti-realism in Mathematics." *Philosophy of Mathematics* (2009): 35–101.
Barad, Karen. *Meeting the Universe Halfway: Quantum Physics and the Entanglement of Matter and Meaning.* Durham: Duke University Press, 2007.
———. "Posthumanist Performativity: Toward an Understanding of How Matter Comes to Matter." *Signs: Journal of Women in Culture and Society* 28 (2003): 801–31.
Benardete, José A. *Infinity: An Essay in Metaphysics.* Oxford: Clarendon Press, 1964.
Benmelech, Efraim, and Jennifer Dlugosz. "The Credit Rating Crisis." *NBER Macroeconomics Annual* 24 (2009): 161–207.
Bennett, Jane. *Vibrant Matter: A Political Ecology of Things.* Durham: Duke University Press, 2010.

Benoit-Barné, Chantal. "Socio-technical Deliberation About Free and Open Source Software: Accounting for the Status of Artifacts in Public Life." *Quarterly Journal of Speech* 93 (2007): 211–35.

Berkeley, George. *The Analyst*. Vol. 3 of *The Works of George Berkeley*. Edited by Alexander Campbell Fraser. Oxford: Clarendon Press, 1901.

———. *The Analyst, or A Discourse Addressed to an Infidel Mathematician*. London: Tonson, 1734.

Bettadpur, Srinivas. "Matter in Motion: Earth's Changing Gravity." NASA EarthData. Accessed June 12, 2020. https://earthdata.nasa.gov/learn/sensing-our-planet/matter-in-motion-earth-s-changing-gravity.

Biesecker, Barbara A., and John Louis Lucaites, eds. *Rhetoric, Materiality, and Politics*. New York: Lang, 2009.

Bishop, Michael. "Semantic Flexibility in Scientific Practice: A Study of Newton's Optics." *Philosophy and Rhetoric* 32 (1999): 210–32.

Blank, Brian. Book Review of *An Imaginary Tale: The Story of $\sqrt{-1}$* by Paul Nahin. *Notices of the AMS* 46 (1999): 1233–36.

Boaler, Jo. "Mathematics from Another World: Traditional Communities and the Alienation of Learners." *Journal of Mathematical Behavior* 18 (2000): 379–97.

Bombelli, Rafael. *L'algebra*. Bologna: Rossi, 1579.

Bombelli, Rafael, and Cataldi. "Bombelli and Cataldi on Continued Fractions." In *A Sourcebook in Mathematics*, translated by V. Sanford, edited by David Eugene Smith, 80–84. New York: Dover, 1959.

Boyer, Carl B. *The History of the Calculus and Its Conceptual Development (the Concepts of the Calculus)*. New York: Dover, 1959.

Boyer, Carl B., and Uta C. Merzbach. *A History of Mathematics*. 2nd ed. Hoboken: Wiley and Sons, 1991.

Brummelen, Glen van. *The Mathematics of the Heavens and the Earth: The Early History of Trigonometry*. Princeton: Princeton University Press, 2009.

Burgess, Neil, Eleanor A. Maguire, and John O'Keefe. "The Human Hippocampus and Spatial and Episodic Memory." *Neuron* 35 (2002): 625–41.

Burke, Kenneth. *A Grammar of Motives*. New York: Prentice Hall, 1945.

———. *A Rhetoric of Motives*. Berkeley: University of California Press, 1969.

Burkert, Walter. *Lore and Science in Ancient Pythagoreanism*. Translated by E. Minar. Cambridge, MA: Harvard University Press, 1972.

Burton, Leone. "The Practices of Mathematicians: What Do They Tell Us About Coming to Know Mathematics?" *Educational Studies in Mathematics* 37 (1998–99): 121–43.

Burtt, Edwin. *The Metaphysical Foundations of Modern Physical Science: A Historical and Critical Essay*. London: Compton, 1967.

Cajori, Florian. Appendix to Newton, *Sir Isaac Newton's Mathematical Principles*, 652–72.

Cardano, Gerolamo. *Ars magna, or The Rules of Algebra*. 1545. Reprint, New York: Dover, 1993.

Cavalieri, Bonaventura. *Geometria degli indivisibili*. Edited by Lucio Lombardo-Radice. Torino: Unione Tipografico–Editrice Torinese, 1966.

———. *Geometria indivisibilibus continuorum nova quadam ratione promota*. Italy: Ex Typographia de Duciis, 1653.

Ceccarelli, Leah. "Rhetorical Criticism and the Rhetoric of Science." *Western Journal of Speech Communication* 65 (2001): 314–29.

Cellucci, Carlo. "Philosophy of Mathematics: Making a Fresh Start." *Studies in History and Philosophy of Science* 44 (2013): 32–42.

Changeux, Jean-Pierre, and Alain Connes. *Conversations on Mind, Matter, and Mathematics*. Translated and edited by M. B. DeBevoise. Princeton: Princeton University Press, 1995.

Chaput, Catherine, and Crystal Broch Colombini. "The Mathematization of the Invisible Hand: Rhetorical Energy and the Crafting of Economic Spontaneity." In Wynn and Reyes, *Arguing with Numbers*, 55–81.
Charland, Maurice. "Constitutive Rhetoric: The Case of the *Peuple Quebecois*." *Quarterly Journal of Speech* 73 (1987): 133–50.
Child, J. M. "Critical and Historical Notes." In Leibniz, *Early Mathematical Manuscripts*, 147–58.
Cifoletti, Giovana. "From Valla to Viète: The Rhetorical Reform of Logic and Its Use in Early Modern Algebra." *Early Science and Medicine* II (2006): 390–423.
———. "Mathematics and Rhetoric: Introduction." *Early Science and Medicine* (2006): 369–89.
Collier, James. "Reclaiming Rhetoric of Science and Technology." *Technical Communication Quarterly* 14 (2005): 295–302.
Condit, Celeste. "Mind the Gaps: Hidden Purposes and Missing Internationalism in Scholarship on the Rhetoric of Science and Technology in Public Discourse." *Poroi* 9 (2013): 1–9.
———. "Race and Genetics from a Modal Materialist Perspective." *Quarterly Journal of Speech* 94 (2008): 383–406.
Coval, Joshua, Jakub Jurek, and Erik Stafford. "The Economics of Structured Finance." *Journal of Economic Perspectives* 23 (2009): 3–25.
Cyranoski, David. "What CRISPR-Baby Prison Sentences Mean for Research." *Nature* 577 (2020): 154–55.
Davis, Diane. "Rhetoricity at the End of the World." *Philosophy and Rhetoric* 50 (2017): 431–51.
Davis, Diane, and Michelle Ballif. "Introduction: Pushing the Limits of the *Anthropos*." Special issue, *Philosophy and Rhetoric* 47 (2014): 346–53.
———, eds. "Pushing the Limits of the *Anthropos*." Special issue, *Philosophy and Rhetoric* 47 (2014).
Davis, Philip J., and Reuben Hersh. *Descartes' Dream: The World According to Mathematics*. San Diego: Harcourt Brace Jovanovich, 1986.
———. *The Mathematical Experience*. Boston: Birkhauser, 1995.
———. "Rhetoric and Mathematics." In *The Rhetoric of the Human Sciences: Language and Argument in Scholarship and Public Affairs*, edited by John S. Nelson, Allan Megill, and Deirdre N. McCloskey, 53–68. Madison: University of Wisconsin Press, 1987.
Deleuze, Gilles. "Postscript on the Societies of Control." *October* 59 (1992): 3–7.
De Morgan, Augustus. "On the Early History of Infinitesimals in England." *London, Edinburgh, and Dublin Philosophical Magazine and Journal of Science* 4 (1852): 321–30.
Depew, David J., and John Lyne. "The Productivity of Scientific Rhetoric." *Poroi* 9 (2013): 1–20. Reprinted in *Landmark Essays on Rhetoric of Science: Issues and Methods*, edited by Randy Allen Harris, 130–44. New York: Routledge, 2020.
Desilet, Gregory. "Physics and Language—Science and Rhetoric: Reviewing the Parallel Evolution of Theory on Motion and Meaning in the Aftermath of the Sokal Hoax." *Quarterly Journal of Speech* 85 (1999): 339–60.
Dyson, Freeman. "Biological and Cultural Evolution: Six Characters in Search of an Author." *Edge*. June 13, 2019. https://www.edge.org/conversation/freeman_dyson-biological-and-cultural-evolution.
Einstein, Albert. *Sidelights on Relativity*. Translated by G. B. Jeffery and W. Perrett. London: Metheun, 1922.
Enriquez, Juan, and Steve Gullans. *Evolving Ourselves: How Unnatural Selection and Nonrandom Mutation Are Changing Life on Earth*. New York: Penguin, 2015.

Ernest, Paul. "Forms of Knowledge in Mathematics and Mathematics Education: Philosophical and Rhetorical Perspectives." *Educational Studies in Mathematics: An International Journal* 38 (1999): 67–83.

———. *Social Constructivism as a Philosophy of Mathematics.* Albany: State University of New York Press, 1998.

Euclid. *The Thirteen Books of Euclid's Elements.* Translated and Edited by T. L. Heath. New York: Dover, 1956.

Eves, Howard. *Great Moments in Mathematics: After 1650.* Washington, DC: Mathematical Association of America, 1981.

Fabozzi, Frank J., Anand K. Bhattacharya, and William S. Berliner. *Mortgage-Backed Securities Products, Structuring, and Analytical Techniques.* 2nd ed. New York: Wiley, 2011.

Fahnestock, Jeanne. "The New Mathematical Arts of Argument: Naturalistic Images and Geometric Diagrams." In Wynn and Reyes, *Arguing with Numbers,* 171–211.

Fausto-Sterling, Anne. "Bare Bones of Race." *Social Studies of Science* 38 (2008): 657–94.

———. "The Bare Bones of Sex: Part 1; Sex and Gender." *Signs: Journal of Women in Culture and Society* 30 (2005): 1491–1527.

Federal Reserve Bank of St. Louis. "Delinquency Rate on Single-Family Residential Mortgages, Booked in Domestic Offices, All Commercial Banks." Federal Reserve Bank of St. Louis Economic Research. August 15, 2017. http://research.stlouisfed.org/fred2/series/drsfrmacbs.

Financial Crisis Inquiry Commission. *The Financial Crisis Inquiry Report: Final Report of the National Commission on the Causes of the Financial and Economic Crisis in the United States.* Official government ed. Washington, DC: Financial Crisis Inquiry Commission, 2011.

Finn, Ed. *What Algorithms Want: Imagination in the Age of Computing.* Cambridge, MA: MIT Press, 2017.

Fontenelle, Bernard de. *Éléments de la géométrie de l'infini.* Paris, 1727.

Foucault, Michel. *Archaeology of Knowledge.* London: Routledge, 2002.

———. *The Order of Things: An Archaeology of the Human Sciences.* London: Routledge, 2002.

Fowler, David H. *The Mathematics of Plato's Academy: A New Reconstruction.* Oxford: Oxford University Press, 1987.

Frege, Gottlob. "On Sense and Reference." In *Translations from the Philosophical Writings of Gottlob Frege.* Translated by Peter Geach and Max Black, 56–78. Oxford: Basil Blackwell, 1960.

———. "The Thought: A Logical Inquiry." Translated by A. M. Quinton and Marcelle Quinton. *Mind* 65 (1956): 289–311.

Fuller, Steve, and James H. Collier. *Philosophy, Rhetoric, and the End of Knowledge: A New Beginning for Science and Technology Studies.* Mahwah, NJ: Erlbaum, 2004.

Gantumur, Tsogtgerel. "Complex Numbers." McGill University. Accessed June 13, 2020. http://www.math.mcgill.ca/gantumur/math249w15/numbers.pdf.

Gaonkar, Dilip. "The Idea of Rhetoric in the Rhetoric of Science." In Gross and Keith, *Rhetorical Hermeneutics,* 25–88. Originally published in *Southern Communication Journal* 58 (1993): 258–95.

Gauss, Carl Friedrich. *Werke.* Vol. 1. Cambridge: Cambridge University Press, 2011.

Gillespie, Tarleton. "Algorithm." In *Digital Keywords: A Vocabulary of Information Society and Culture,* edited by Ben Peters, 18–30. Princeton: Princeton University Press, 2016.

———. "The Relevance of Algorithms." In *Media Technologies: Essays on Communication, Materiality, and Society,* edited by Tarleton Gillespie, Pablo Boczkowski, and Kirsten Foot, 167–94. Cambridge, MA: MIT Press, 2014.

Gödel, Kurt Friedrich. *Collected Works.* Vol. 2, *Publications, 1938–1974*, edited by Solomon Feferman, John W. Dawson Jr., Stephen C. Kleene, Gregory H. Moore, Robert M. Solovay, and Jean van Heijenoort. Oxford: Oxford University Press, 1995.
Goodnight, Thomas G., and Sandy Green. "Rhetoric, Risk, and Markets: The Dot-Com Bubble." *Quarterly Journal of Speech* 96 (2010): 115–40.
Gould, Stephen Jay. *The Mismeasure of Man.* New York: Norton, 1996.
Graham, S. Scott. *Where's the Rhetoric? Imagining a Unified Field.* Columbus: Ohio State University Press, 2020.
Greene, Ronald Walter. "Rhetoric and Capitalism: Rhetorical Agency as Communicative Labor." *Philosophy and Rhetoric* 37 (2004): 188–206.
Gross, Alan G. "Alan Gross in His Own Words: An Interview in the Association of Rhetoric of Science and Technology Oral History Project." *Poroi* 10 (2014): https://doi.org/10.13008/2151-2957.1212.
———. "ARST Oral History Project: An Interview with Dr. Alan Gross." YouTube video, 32:04. Accessed November 8, 2021. https://www.youtube.com/watch?v=KXCZiYVroLc.
———. *Starring the Text: The Place of Rhetoric in Science Studies.* Carbondale: Southern Illinois University Press, 2006.
Gross, Alan G., and William M. Keith, eds. *Rhetorical Hermeneutics: Invention and Interpretation in the Age of Science.* Albany: State University of New York Press, 1997.
Guthrie, W. K. *A History of Greek Philosophy: Earlier Presocratics and the Pythagoreans.* Cambridge: Cambridge University Press, 1979.
Hacking, Ian. *Why Is There Philosophy of Mathematics at All?* Cambridge: Cambridge University Press, 2014.
———. "Why Is There Philosophy of Mathematics at All?" *South African Journal of Philosophy* 30 (2011): 1–15.
Hall, Alfred R. *Philosophers at War.* Cambridge: Cambridge University Press, 2009.
Hanan, Joshua S. "Home Is Where the Capital Is: The Culture of Real Estate in an Era of Control Societies." *Communication and Critical/Cultural Studies* 7 (2010): 176–201.
Hanan, Joshua S., Indradeep Ghosh, and Kaleb W. Brooks. "Banking on the Present: The Ontological Rhetoric of Neo-classical Economics and Its Relation to the 2008 Financial Crisis." *Quarterly Journal of Speech* 100 (2014): 139–62.
Hanan, Joshua S., and Mark Hayward, eds. *Communication and the Economy: History, Value, and Agency.* New York: Lang, 2014.
Haraway, Donna J. *Modest_Witness@Second_Millennium. FemaleMan_Meets_OncoMouse: Feminism and Technoscience.* 2nd ed. New York: Routledge, 2018.
———. *Primate Visions: Gender, Race, and Nature in the World of Modern Science.* New York: Routledge, 2015.
———. "The Promises of Monsters: A Regenerative Politics for Inappropriate/d Others." In *Cultural Studies*, edited by Lawrence Grossberg, Cory Nelson, and Paula Treichler, 295–337. New York: Routledge, 1992.
———. *Simians, Cyborgs, and Women: The Reinvention of Nature.* New York: Routledge, 2015.
Hardt, Michael, and Antonio Negri. *Empire.* Cambridge, MA: Harvard University Press, 2000.
Hariman, Robert. *Political Style: The Artistry of Power.* Chicago: University of Chicago Press, 1995.
Haug, Espen Gaarder, and Nassim Nicholas Taleb. "Option Traders Use (Very) Sophisticated Heuristics, Never the Black-Scholes-Merton Formula." *Journal of Economic Behavior and Organization* 77 (2011): 97–106.

Havelock, Eric A. "The Linguistic Task of the Presocratics." In *Language and Thought in Early Greek Philosophy*, edited by Kevin Robb, 7–41. La Salle, IL: Monist Library of Philosophy, 1983.
Heath, Thomas Little. *Diaphantus of Alexandria: A Study in Greek Algebra*. Cambridge: Cambridge University Press, 1910.
———. *A History of Greek Mathematics*. 4 vols. New York: Dover, 1981.
Heidegger, Martin. "Modern Science, Metaphysics, and Mathematics." In *Basic Writings: From "Being and Time" (1927) to the "Task of Thinking" (1964)*, translated by W. B. Barton Jr. and Vera Deutsch, edited by David Farrell Krell, 261–300. San Francisco: HarperCollins, 1977.
Helbing, Dirk, Bruno S. Frey, Gerd Gigerenzer, Ernst Hafen, Michael Hagner, Yvonee Hofstetter, Jeroen van den Hoven, Roberto V. Zicari, and Andrej Zwitter. "Will Democracy Survive Big Data and Artificial Intelligence?" *Scientific American*, February 25, 2017. https://www.scientificamerican.com/article/will-democracy-survive-big-data-and-artificial-intelligence/?print=true.
Hill, Kashmir. "Wrongfully Accused by an Algorithm." *New York Times*, June 24, 2020. https://www.nytimes.com/2020/06/24/technology/facial-recognition-arrest.html.
Ho, K. F., and John G. Hedberg. "Teachers' Pedagogies and Their Impact on Students' Mathematical Problem Solving." *Journal of Mathematical Behavior* 24 (2005): 238–52.
Horky, Phillip Sidney. *Plato and Pythagoreanism*. Oxford: Oxford University Press, 2013.
Houck, Davis W. *Rhetoric as Currency: Hoover, Roosevelt, and the Great Depression*. College Station: Texas A&M Press, 2001.
International Monetary Fund. "Historical GDP by Country, 1960–2019." International Monetary Fund Research. July 14, 2017. https://knoema.com/mhrzolg/historical-gdp-by-country-statistics-from-the-world-bank-1960-2019.
Ishino, Yoshizumi, Mart Krupovic, and Patrick Forterre. "History of CRISPR-Cas from Encounter with a Mysterious Repeated Sequence to Genome Editing Technology." *Journal of Bacteriology* 200 (2018): e00580-17. https://doi.org/10.1128/JB.00580-17.
Isocrates. *On the Peace. Areopagiticus. Against the Sophists. Antidosis. Panathenaicus*. Translated by George Norlin. Cambridge, MA: Harvard University Press, 1979.
Issacharoff, Samuel. "Gerrymandering and Political Cartels." *Harvard Law Review* 116 (2002): 593–648.
Jack, Jordynn. *Science on the Home Front: American Women Scientists in World War II*. Urbana: University of Illinois Press, 2009.
Jesseph, Douglas M. "Philosophical Theory and Mathematical Practice in the Seventeenth Century." *Studies in History and Philosophy of Science* 20 (1989): 215–44.
Johnson, Neil, Guannan Zhao, Eric Hunsader, Hong Qi, Nicholas Johnson, Jing Meng, and Brian Tivnan. "Abrupt Rise of New Machine Ecology Beyond Human Response Time." *Scientific Reports* 3 (2013): 2627.
Jones, Sam. "The Formula That Felled Wall St." *Financial Times*, April 24, 2009. https://www.ft.com/content/912d85e8-2d75-11de-9eba-00144feabdc0.
Kahneman, Daniel. *Thinking Fast and Slow*. New York: Farrar, Straus, and Giroux, 2011.
Kennedy, George. "A Hoot in the Dark: The Evolution of General Rhetoric." *Philosophy and Rhetoric* 25 (1992): 1–21.
Kennedy, J. B. "Plato's Forms, Pythagorean Mathematics, and Stichometry." *Aperion* 43 (2010): 1–32.
Kennedy, Sheila Suess. "Electoral Integrity: How Gerrymandering Matters." *Public Integrity* 19 no. 3 (2017): 265–73.

Kline, Morris. *Mathematical Thought from Ancient to Modern Time*. New York: Oxford University Press, 1972.
Kockelman, Paul. "The Anthropology of an Equation: Sieves, Spam Filters, Agentive Algorithms, and Ontologies of Transformation." *HAU: Journal of Ethnographic Theory* 3 (2013): 33–61.
Krugman, Paul. *End This Depression Now*. New York: Norton, 2013.
———. *The Return of Depression Economics and the Crisis of 2008*. New York: Norton, 2009.
Kuhn, Thomas S. *The Structure of Scientific Revolutions*. 2nd ed. Chicago: University of Chicago Press, 1970.
Kurzweil, Ray. *The Singularity Is Near: When Humans Transcend Biology*. New York: Viking, 2005.
Lakatos, Imre. *Proofs and Refutations: The Logic of Mathematical Discovery*. Edited by John Worrall and Elie Zahar. Cambridge: Cambridge University Press, 1976.
Lakoff, George, and Rafael Núñez. *Where Mathematics Comes From: How the Embodied Mind Brings Mathematics into Being*. New York: Basic Books, 2000.
Latour, Bruno. *Pandora's Hope: Essays on the Reality of Science Studies*. Cambridge, MA: Harvard University Press, 1999.
———. *The Pasteurization of France*. Cambridge, MA: Harvard University Press, 1988.
———. *Reassembling the Social: An Introduction to Actor-Network-Theory*. Oxford: Oxford University Press, 2005.
———. *Science in Action: How to Follow Scientists and Engineers Through Society*. Cambridge, MA: Harvard University Press, 1987.
———. *We Have Never Been Modern*. Translated by Catherine Porter. Cambridge, MA: Harvard University Press, 1993.
Lattis, James A. *Between Copernicus and Galileo: Christopher Clavius and the Collapse of Ptolemaic Astronomy*. Chicago: University of Chicago Press, 1994.
Ledford, Heidi. "Quest to Use CRISPR Against Disease Gains Ground." *Nature* 577 (2020): 156.
Leibniz, Gottfried Wilhelm. *The Early Mathematical Manuscripts of Leibniz*. Translated and edited by J. M. Child. Chicago: Open Court, 1920.
———. *Leibniz Selections*. Edited by Philip P. Wiener. New York: Scribner's Sons, 1979.
———. "A New Method for *Maxima* and *Minima* as Well as Tangents, Which Is Neither Impeded by Fractional nor Irrational Quantities, and a Remarkable Type of Calculus for Them." In *A Source Book in Mathematics, 1200–1800*, edited by Dirk Jan Struik, 271–80. Cambridge, MA: Harvard University Press, 1969.
———. *Opera Omnia*. 6 Vols. Edited by Louis Dutens. Geneva: Hildesheim, 1768.
———. *Philosophische schriften*. 7 vols. Berlin: Akademie, 1999.
Lenglet, Marc. "Algorithms and the Manufacture of Financial Reality." In *Objects and Materials*, edited by Penny Harvey, Eleanor Conlin Casella, Gillian Evans, Hannah Knox, Christine McLean, Elizabeth B. Silva, Nicholas Thoburn, and Kath Woodward, 312–22. London: Routledge, 2013.
Lewis, Michael. *The Big Short: Inside the Doomsday Machine*. New York: Norton, 2010.
Li, David. "On Default Correlation: A Copula Function Approach." SSRN. August 15, 2017. http://papers.ssrn.com/sol3/papers.cfm?abstract_id=187289, 1–20. Originally published in *Journal of Fixed Income* 9 (2000): 43–54.
Little, Joseph. "Rhetoric and Mathematics in the Saturnian Account of Atomic Spectra." In Wynn and Reyes, *Arguing with Numbers*, 151–70.
Luce, Arthur A. *Berkeley and Malebranche: A Study in the Origins of Berkeley's Thought*. London: Oxford University Press, 1934.

Lynch, Paul, and Nathaniel Rivers, eds. *Thinking with Bruno Latour in Rhetoric and Composition.* Carbondale: Southern Illinois University Press, 2015.
MacKenzie, Donald. *An Engine, Not a Camera: How Financial Models Shape Markets.* Cambridge, MA: MIT Press, 2006.
Madrigal, Alexis C. "How Netflix Reverse-Engineered Hollywood." *Atlantic,* January 2, 2014. http://theatlantic.com/technology/archive/2014/01/how-netflix-reverse-engineered-hollywood/282679.
Maki, Uskali. "How to Combine Rhetoric and Realism in the Methodology of Economics." *Economics and Philosophy* 4 (1988): 89–109.
Mbembé, Achille, and Steve Corcoran. *Necropolitics.* Durham: Duke University Press, 2019.
Mbembé, Achille, and Laurent Dubois. *Critique of Black Reason.* Durham: Duke University Press, 2017.
McCloskey, Deidre. *The Rhetoric of Economics.* 2nd ed. Madison: University of Wisconsin Press, 1998.
———. "Two Replies and a Dialogue on the Rhetoric of Economics: Maki, Rappaport, and Rosenberg." *Economics and Philosophy* 4 (1988): 150–66.
McLean, Bethany, and Joseph Nocera. *All the Devils Are Here: The Hidden History of the Financial Crisis.* New York: Portfolio/Penguin, 2011.
Merriam, Allen H. "Words and Numbers: Mathematical Dimensions of Rhetoric." *Southern Communication Journal* 55 (1990): 337–54.
Miller, Carolyn R. "Novelty and Heresy in the Debate on Nonthermal Effects of Electromagnetic Fields." In *Rhetoric and Incommensurability,* edited by Randy Allen Harris, 464–506. West Lafayette: Parlor Press, 2005.
———. "The Presumption of Expertise: The Role of Ethos in Risk Analysis." *Configurations* 11 (2003): 163–202.
Mirowski, Philip. *More Heat Than Light: Economics as Social Physics, Physics as Nature's Economics.* Cambridge: Cambridge University Press, 1991.
———. "Shall I Compare Thee to a Minkowski-Ricardo-Leontief-Metzler Matrix of the Mosak-Hicks Type? Or, Rhetoric, Mathematics, and the Nature of Neoclassical Economic Theory." In *The Consequences of Economic Rhetoric,* edited by Arjo Klamer, Donald N. McCloskey, and Robert M. Solow, 117–45. Cambridge: Cambridge University Press, 1988.
Mittelstadt, Brent Daniel, Patrick Allo, Mariarosaria Taddeo, Sandra Wachter, and Luciano Floridi. "The Ethics of Algorithms: Mapping the Debate." *Big Data and Society* 3 (2016): 1–21.
Moravec, Eva Ruth. "Do Algorithms Have a Place in Policing?" *Atlantic,* September 5, 2019. https://www.theatlantic.com/politics/archive/2019/09/do-algorithms-have-place-policing/596851.
Morgenson, Gretchen, and Joshua Rosner. *Reckless Endangerment: How Outsized Ambition, Greed, and Corruption Led to Economic Armageddon.* New York: Times Books/Holt, 2011.
Mudry, Jessica J. *Measured Meals: Nutrition in America.* Albany: State University of New York Press, 2009.
Nahin, Paul J. *An Imaginary Tale: The Story of [the Square Root of Minus One].* Princeton: Princeton University Press, 1998.
Netz, Reviel. *The Shaping of Deduction in Greek Mathematics: A Study in Cognitive History.* Cambridge: Cambridge University Press, 1999.
Newton, Isaac. *Isaaci Newtoni opera quae exstant omnia.* 5 vols. Edited by Samuel Horsley. London, 1779–85.

———. *The Method of Fluxions and Infinite Series: With Its Application to the Geometry of Curve-Lines*. Edited by John Colson. London: Woodfall/Nourse, 1736.

———. *Sir Isaac Newton's Mathematical Principles of Natural Philosophy and His System of the World*. Translated by Andrew Motte. Edited by Florian Cajori. Berkeley: University of California Press, 1962.

Newton, Isaac, and I. Bernard Cohen. *Isaac Newton's Papers and Letters on Natural Philosophy and Related Documents*. Edited by I. Bernard Cohen. Cambridge, MA: Harvard University Press, 1978.

Neyland, Daniel. "On Organizing Algorithms." *Theory, Culture, and Society* 32 (2015): 119–32.

Nietzsche, Friedrich Wilhelm. *Philosophy and Truth: Selections from Nietzsche's Notebooks of the Early 1870s*. Translated and edited by Daniel Breazeale. Amherst: Humanity Books, 1999.

———. *The Philosophy of Nietzsche*. Translated by Thomas Common, Helen Zimmern, Horace Barnett Samuel, J. M. Kennedy, and Clifton Fadiman. New York: Modern Library, 1927.

Nussbaum, Martha. *The Fragility of Goodness: Luck and Ethics in Greek Tragedy and Philosophy*. Cambridge: Cambridge University Press, 2001.

O'Meara, Dominic J. *Pythagoras Revived: Mathematics and Philosophy in Late Antiquity*. New York: Oxford University Press, 1989.

O'Neil, Cathy. *Weapons of Math Destruction: How Big Data Increases Inequality and Threatens Democracy*. New York: Crown, 2016.

Pasquale, Frank. *The Black Box Society: The Secret Algorithms That Control Money and Information*. Cambridge, MA: Harvard University Press, 2015.

Perelman, Chaim. *The Realm of Rhetoric*. Translated by William Kluback. Notre Dame: University of Notre Dame Press, 1982.

Perelman, Chaim, and Lucie Olbrechts-Tyteca. *The New Rhetoric: A Treatise on Argumentation*. Translated by Johan Wilkinson and Purcell Weaver. Notre Dame: University of Notre Dame Press, 1969.

Philip, J. A. *Pythagoras and Early Pythagoreans*. Toronto: University of Toronto Press, 1966.

Plato. *Cratylus*. Vol. 4 of *Plato*. Translated by Harold North Fowler, W. R. M. Lamb, and Robert Gregg Bury. Loeb Classical Library. Cambridge, MA: Harvard University Press; London: Heinemann, 1928.

———. *Euthyphro, Apology, Crito, Phaedo*. Edited and Translated by Chris Emlyn-Jones and William Preddy. Loeb Classical Library. Cambridge, MA: Harvard University Press, 2017.

———. *Laches, Protagoras, Meno, Euthydemus*. Vol. 2 of *Plato*. Translated by Harold North Fowler, W. R. M. Lamb, and Robert Gregg Bury. Loeb Classical Library. Cambridge, MA: Harvard University Press; London: Heinemann, 1928.

———. *Lysis, Symposium, Gorgias*. Translated by W. R. M. Lamb. Loeb Classical Library. Cambridge, MA: Harvard University Press, 1996.

———. *Plato: Complete Works*. Edited by John M. Cooper and D. S. Hutchinson. Translated by R. D. McKirahan. Indianapolis: Hackett, 1997.

———. *Republic*. Vols. 5–6 of *Plato*. Translated by Harold North Fowler, W. R. M. Lamb, and Robert Gregg Bury. Loeb Classical Library. Cambridge, MA: Harvard University Press; London: Heinemann, 1928.

———. *The Republic of Plato*. Translated by Alan Bloom. Basic Books, 1968.

———. *Theaetetus, Sophist*. Vol. 7 of *Plato*. Translated by Harold North Fowler. Loeb Classical Library. Cambridge, MA: Harvard University Press, 1996.

———. *Timaeus, Critias, Cleitophon, Menexenus, Epistles*. Translated by Robert Gregg Bury. Loeb Classical Library. Cambridge, MA: Harvard University Press, 1981.
Plutarch. *Lives, Volume V, Marcellus*. Translated by Bernadotte Perrin. Loeb Classical Library. Cambridge, MA: Harvard University Press, 1968.
Porter, Theodore M. *Trust in Numbers: The Pursuit of Objectivity in Science and Public Life*. Princeton: Princeton University Press, 1995.
Prenosil, Joshua. "Bruno Latour Is a Rhetorician of Inartistic Proofs." In Lynch and Rivers, *Thinking with Bruno Latour*, 97–114.
Putnam, Hilary. *Mathematics, Matter, and Method*. 2nd ed. Cambridge: Cambridge University Press, 1979.
Raphson, Joseph. *The History of Fluxions, Showing in a Compendious Manner the First Rise of, and Various Improvements Made in That Incomparable Method*. London, 1715.
Rav, Yehuda. "A Critique of a Formalist-Mechanist Version of the Justification of Arguments in Mathematicians' Proof Practices." *Philosophia Mathematica* 15 (2007): 291–332.
Reid, Rob. "AI Boom—or Doom?" *After On*. Podcast, episode 47, 1:30:06. April 30, 2019. https://after-on.com/episodes-31-60/047.
Resnik, Michael. *Mathematics as a Science of Patterns*. Oxford: Oxford University Press, 1997.
Reyes, G. Mitchell. "Algorithms and Rhetorical Inquiry: The Case of the 2008 Financial Collapse." *Rhetoric and Public Affairs* 22 (2019): 569–614.
———. "The Horizons of Judgement in Mathematical Discourse: Copulas, Economics, and Subprime Mortgages." In Wynn and Reyes, *Arguing with Numbers*, 82–121.
———. "The Rhetoric in Mathematics: Newton, Leibniz, the Calculus, and the Rhetorical Force of the Infinitesimal." *Quarterly Journal of Speech* 90 (2004): 163–88. https://www.tandfonline.com/doi/full/10.1080/0033563042000227427.
———. "Stranger Relations: The Case for Rebuilding Commonplaces Between Rhetoric and Mathematics." *Rhetoric Society Quarterly* 44 (2014): 470–91.
———. "The Swift Boat Veterans for Truth, the Politics of Realism, and the Manipulation of Vietnam Remembrance in the 2004 Presidential Election." *Rhetoric and Public Affairs* 9 (2006): 571–600.
Reyes, G. Mitchell, and Kundai Chirindo. "Theorizing Race and Gender in the Anthropocene." *Women's Studies in Communication* 43 (2020): 429–42.
Richland, Lindsey E., Keith J. Holyoak, and James W. Stigler. "Analogy Use in Eighth-Grade Mathematics Classrooms." *Cognition and Instruction* 22 (2004): 37–60.
Rickert, Thomas J. *Ambient Rhetoric: The Attunements of Rhetorical Being*. Pittsburgh: University of Pittsburgh Press, 2013.
Riedweg, Christoph. *Pythagoras: His Life, Teaching and Influence*. Translated by Steven Rendall. Ithaca: Cornell University Press, 2005.
Robinson, Abraham. *Nonstandard Analysis*. Princeton: Princeton University Press, 1996.
Rorty, Richard. *Contingency, Irony, Solidarity*. Cambridge: Cambridge University Press, 1989.
Rotman, Brian. *Ad Infinitum: The Ghost in Turing's Machine; Taking God Out of Mathematics and Putting the Body Back In*. Stanford: Stanford University Press, 1993.
———. *Mathematics as Sign: Writing, Imagining, Counting*. Stanford: Stanford University Press, 2000.
———. *Signifying Nothing: The Semiotics of Zero*. New York: St. Martin's Press, 1987.
Rouse, Joseph. *Engaging Science: How to Understand Its Practices Philosophically*. Ithaca: Cornell University Press, 1996.
Russell, Bertrand. *Power: A New Social Analysis*. London: Allen and Unwin, 1938.
———. "The Study of Mathematics." *New Quarterly* 1 (1907): 30–42.
Russell, Stuart. *Human-Compatible Artificial Intelligence*. New York: Viking, 2019.

Russell, Stuart, Daniel Dewey, and Max Tegmark. "Research Priorities for Robust and Beneficial Artificial Intelligence." *AI Magazine* 36 (2015): 105–14.
Saffrey, H. D. "*Ageômetrêtos mêdeis eisitô*: Une inscription légendaire." *Revue des Études Grecques* 81 (1968): 67–87.
———. *Recherches sur le néoplatonisme après Plotin*. Paris: Vrin, 1990.
Salmon, Felix. "The Formula That Killed Wall Street." *Significance* 9 (2012): 16–20.
Sautoy, Marcus du, and Caroline Series. "In Our Time: Imaginary Numbers." BBC Sounds. Accessed June 12, 2020. https://www.bbc.co.uk/sounds/play/b00tt6b2.
Schiappa, Edward. "Did Plato Coin Rhêtorikê?" *American Journal of Philology* 111 (1990): 457–70.
———. "In What Ways Shall We Describe Mathematics as Rhetorical?" In Wynn and Reyes, *Arguing with Numbers*, 33–52.
Schiralli, Martin, and Nathalie Sinclair. "A Constructive Response to 'Where Mathematics Comes From.'" *Educational Studies in Mathematics* 52 (2003): 79–91.
Serres, Michel. "Gnomon: The Beginnings of Geometry in Greece." In *A History of Scientific Thought: Elements of a History of Science*, edited by Michel Serres, 73–123. Oxford: Blackwell, 1995.
Simonson, Peter. "Review Essay: Rhetoric Culture, Things." *Quarterly Journal of Speech* 100 (2014): 105–25.
Singh, Simon. "The Imaginary Number." BBC Science. Accessed June 13, 2020. http://www.bbc.co.uk/radio4/science/5numbers4.shtml.
Smith, Andrew. "Franken-Algorithms: The Deadly Consequences of Unpredictable Code." *Guardian*, August 30, 2018. https://www.theguardian.com/technology/2018/aug/29/coding-algorithms-frankenalgos-program-danger.
Social Media Collective. "Critical Algorithm Studies." Accessed November 8, 2021. https://socialmediacollective.org/reading-lists/critical-algorithm-studies.
Sparrow, Bety, Jenny Liu, and Daniel M. Wegner. "Google Effects on Memory: Cognitive Consequences of Having Information at Our Fingertips." *Science* 333 (2011): 776–78.
Stewart, Ian. "Sweet Nothings." *New Scientist* 173 (2002): 27–28.
Stiegler, Bernard. *Technics and Time*. Vol. 2, *Disorientation*. Stanford: Stanford University Press, 2009.
Stockman, David A. *The Great Deformation: The Corruption of Capitalism in America*. New York: Public Affairs, 2014.
Stormer, Nathan. "Articulation: A Working Paper on Rhetoric and *Taxis*." *Quarterly Journal of Speech* 90 (2004): 257–84.
———. "Rhetoric's Diverse Materiality: Polythetic Ontology and Genealogy." *Review of Communication* 16 (2016): 299–316.
Stormer, Nathan, and Bridie McGreavy. "Thinking Ecologically About Rhetoric's Ontology: Capacity, Vulnerability, and Resilience." *Philosophy and Rhetoric* 50 (2017): 1–25.
Strogatz, Steven. *The Joy of X: A Guided Tour of Math, from One to Infinity*. Boston: Houghton Mifflin, 2012.
Taleb, Nassim. *The Black Swan: The Impact of the Highly Improbable*. New York: Random House, 2007.
Thomas, Neal. "Social Computing as a Platform for Memory." *Culture Machine* 14 (2013): 1–16.
Thurston, William. "On Proof and Progress in Mathematics." *Bulletin of the American Mathematical Society* 30 (1994): 161–77.
Tversky, Amos, and Daniel Kahneman. "Judgment Under Uncertainty: Heuristics and Biases." *Science* 185 (1974): 1124–31.

Tyson, Neil DeGrasse. *Cosmos: A Spacetime Odyssey*. Beverly Hills: Twentieth-Century Fox Home Entertainment, 2014.
Ullman, Ellen. "Programming the Post-human: Computer Science Redefines 'Life.'" *Harper's* 305 (2002): 60–70.
Upchurch, Tom. "To Work for Society, Data Scientists Need a Hippocratic Oath with Teeth." *Wired*. Accessed November 8, 2021. https://www.wired.co.uk/article/data-ai-ethics-hippocratic-oath-cathy-o-neil-weapons-of-math-destruction.
US Technology Policy Committee. "Statement on Principles and Prerequisites for the Development, Evaluation, and Use of Unbiased Facial Recognition Technologies." ACM. July 6, 2020. https://www.acm.org/binaries/content/assets/public-policy/ustpc-facial-recognition-tech-statement.pdf.
Vilenkin, N. Ya. *In Search of Infinity*. Translated by Abe Shenitzer. Boston: Birkhauser, 1995.
Vivian, Bradford. *Being Made Strange: Rhetoric Beyond Representation*. Albany: State University of New York Press, 2004.
Vlastos, Gregory. "Elenchus and Mathematics: A Turning-Point in Plato's Philosophical Development." *American Journal of Philology* 109 (1988): 362–96.
Waerden, B. L. van der, *A History of Algebra: From al-Khwārizmī to Emmy Noether*. Berlin: Springer-Verlag, 1985.
Wallis, John. *A Treatise of Algebra*. London: Playford, 1685.
Walsh, Lynda, Nathaniel A. Rivers, Jenny Rice, Laurie E. Gries, Jennifer L. Bay, Thomas Rickert, and Carolyn R. Miller. "Forum: Bruno Latour on Rhetoric." *Rhetoric Society Quarterly* 47 (2017): 403–62.
Wasserman, David. "Hating Gerrymandering Is Easy. Fixing It Is Harder." *FiveThirtyEight: The Gerrymandering Project*. January 25, 2018. https://fivethirtyeight.com/features/hating-gerrymandering-is-easy-fixing-it-is-harder.
Wessel, Caspar. *On the Analytical Representation of Direction: An Attempt Applied Chiefly to Solving Plane and Spherical Polygons*. Translated by Flemming Damhus. Edited by Bodil Branner and Jesper Lutzen. 1797. Reprint, Copenhagen: Royal Danish Academy of Sciences and Letters, 1999.
Wigner, Eugene P. "The Unreasonable Effectiveness of Mathematics in the Natural Sciences: Richard Courant Lecture in Mathematical Sciences Delivered at New York University, May 11, 1959." *Communications on Pure and Applied Mathematics* 13 (1960): 1–14.
Wright, Crispin. *Realism, Meaning, and Truth*. Oxford: Basil Blackwell, 1987.
Wynn, James, and G. Mitchell Reyes, eds. *Arguing with Numbers: The Intersections of Rhetoric and Mathematics*. University Park: Pennsylvania State University Press, 2020.
———. "From Division to Multiplication: Uncovering the Relationship Between Mathematics and Rhetoric Through Transdisciplinary Scholarship." In Wynn and Reyes, *Arguing with Numbers*, 11–32.
Zalasiewicz, Jan, Mark Williams, Will Steffen, and Paul Crutzen. "The New World of the Anthropocene." *Environmental Science and Technology* 44 (2010): 2228–31.
Zhmud, Leonid. *Pythagoras and Early Pythagoreanism*. Translated by Kevin Windle and Rosh Ireland. Oxford: Oxford University Press, 2012.
Zuboff, Shoshana. *The Age of Surveillance Capitalism: The Fight for a Human Future at the New Frontier of Power*. New York: Public Affairs, 2019.

Index

Italicized page references indicate illustrations.

acceleration, 61
action
    vs. motion, 153n34
addressivity, 127
agency, 8–9, 109
Agent
    relationship between Subject and, 39, 40, 41
aletheia, 59
algebraic trigonometry, 93
algorithms
    agency of, 109, 128
    constitutive approach to, 109, 165n22
    definition of, 128
    development of, 141–42
    displacement of practical judgment by, 129
    of DNA editing, 2
    in election technologies, 1, 2
    ethics of, 107
    fake news and, 107
    financial crisis of 2008 and, 10
    horizons of judgment within, 109, 110
    human-centric approach to, 105, 165n19, 165n24
    implementation of, 103, 107–8
    influence of, 103–4
    monetization of, 130
    one-way mirror metaphor of, 107
    ontological force of, 127
    as principles of composition, 104–5, 116, 123, 124
    problem-solving capabilities, 119, 128
    realities and, 109
    sociocultural consequences of, 106–8, 165n6
    surveillance technology and, 1, 106, 107
    terraforming power of, 120
    translative rhetorical force of, 103
    See also complex algorithms
Amazon, 120
analytic geometry, 93
anchoring effects, 168n72
angles
    numerical signifier of, 93

Anthropocene, 10, 144, 145, 150
Archimedes
    approximation of π, 17
    discoveries of, 47–49
    mathematics of ratios of, 50
Archytas, 21
Argand, Jean-Robert, 96
Argand diagram, 96–98, *97*, *98*, 101
Aristotle
    idea of Form-Numbers, 24
    Metaphysics, 24
    theory of rhetoric, 4, 131, 132
Arkeryd, Leif, 162n81
artificial intelligence, 106
Asimov, Isaac, 3, 5
atomistic theory, 170n32

Bacon, Francis, 80
Balaguer, Mark, 35
Ballif, Michelle, 8, 153n34
Barad, Karen, 50, 54, 145, 157n47, 169n10, 170n32
    Meeting the Universe Halfway, 130, 131
Bay, Jennifer, 141
Benardete, José, 77
Bennett, Jane, 147
Bentley, Richard, 80
Berkeley, George, 66, 80, 159n20, 159n27, 160n42, 161n44
Bettadpur, Srinivas, 170n30
biological evolution, 149
Bishop, Michael, 67
Blank, Brian, 91
Bohr, Neils, 170n32
Boltzmann equation, 162n81
Bombelli, Rafael, 90, 95
Boyer, Carl, 12, 91, 96, 97
Brook, Kaleb W., 164n4
    "Banking on the Present," 166n33
Burke, Kenneth, 65, 153n34
Burton, Leone, 35

186 INDEX

Calculus
    complex curvatures in, 61–62, 62, 158n3, 161n47
    definition of, 61
    emergence of, 8–9, 60–61, 66–67, 81
    Euclidean geometry and, 85
    impact on mathematical discourse, 78–79, 81, 84, 146
    infinitesimals and, 66–69, 78–79, 161n47
    invention of, 132, 146, 158n1
    prediction of motion, 61, 77, 78, 83
    rates of change, 153n36
    as rhetorical practice, 62–63, 65, 80, 82
Cardano, Gerolamo
    Ars magna, 89, 90
    introduction of complex numbers, 90
    solutions of cubic equation, 89, 90
    use of imaginaries, 95
Cartesian coordinate system
    basic element of, 98
    impact on cartography, 51–52, 170n25
    limits of, 99
    negative numbers in, 45–46, 46, 87–88, 88, 99
casus irreducibilis, 90
Cauchy, Augustin-Louis, 77, 79, 83
Ceccarelli, Leah, 140
Changeux, Jean-Pierre, 12
Charland, Maurice, 157n47
classical rhetoric, 169n15
Clavius, Cristoph, 160n33
codes, 141
collateralized debt obligations (CDOs)
    Li copula and, 167n56
    market and, 111, 121–22
    retranching of, 167n60
Collier, James, 170n21
complex algorithms, 165n22, 165n24
complex functions, 100
complex numbers, 90, 91, 96, 97, 98, 100, 163n28
concept-stretching
    idea of, 163n19
conceptual metaphor, 45
Condit, Celeste, 147
congealing of agency, 109
Connes, Alain, 12, 153n4
constitutive rhetoric, 54–55, 82, 85
constructive rhetoric, 65
consummate principle of composition, 100
continuous motion, 61–62, 67, 68, 71, 77
Coval, Joshua, 166n36
creationism, 169n20

credit default swaps (CDSs), 116, 124, 126
Crick, Francis, 99
CRISPR-CAS9 technology, 2
critical methodology, 169n15
cryptocurrencies, 107
cubic equation, 89–90
culture
    evolution of, 149, 171n40
    vs. nature, 144, 148, 152n15

Davis, Diane, 8, 153n34
Davis, Philip, 6, 36, 63–64, 65, 82, 162n71
debt obligations, 118
deductive method, 3
default correlation
    algorithm of, 119
    alternative framework for, 114–15
    approximation of, 112–13, 114
    data for, 116, 119
    formula of, 118
    Li copula and, 113, 119
    between loans, 118
    mathematics of, 166n34
    prediction of, 115
    problem of, 112, 113
    reduction to a single constant, 119, 120, 126
    scalability of, 116
    time dependence of, 114
default risk, 111
democracy
    geometry and, 32
    rhetorical forms of, 34
De Morgan, Augustus, 70
Depew, David J., 169n13, 169n20
derivative, 6
Derrida, Jacques, 3
Descartes, René, 45
    Discourse on the Method, 91
    idea of imaginaries and, 89, 91
    interpretation of negative numbers, 87
    La géométrie, 75, 93
    philosophy of, 99
Desilet, Gregory, 62
diagrams, 19–20, 29–30
dialectics, 18
diffraction, 50, 139, 144
digital pedagogy platforms, 107
Diophantus
    Arithmetica, 87
Duffie, Darrell, 111
Du Sautoy, Marcus, 100
Dyson, Freeman, 149

e (mathematical constant), 91, 94
economics
    mathematization of, 104, 129, 142, 164nn4–5, 167n41
    realist approach to, 166n38
Einstein, Albert, 12
empiricism, 81
eristic rhetoric, 33
Euclid
    axioms, 3
    Elements, 8–9, 64
    on ratio, 64–65
    view of infinitesimals, 64
Euclidean geometry
    Calculus and, 85
    description of motion, 82
    diagrammatic paradigm of, 89
    exclusion of infinitesimals from, 64, 68
    influence of, 60, 160n33
    method of exhaustion, 158n3
    representation in, 81
    rigor in, 161n47
Euler, Leonhard, 91, 95–96, 101
Euler's identity (equation), 91–92, 94–95, 96, 122, 156n30
Euler's number. See *e* (mathematical constant)
Eves, Howard, 61
exponential logarithms, 94–95

Facebook, 105, 130
facial-recognition technology, 1, 2, 151n4
fake news, 107
Fausto-Sterling, Anne, 147
Feferman, Soloman, 162n71
Ferro, Scipione del, 90
Financial Crisis Inquiry Commission, 167n60
financial crisis of 2008, 104, 123
financial instruments, 113–14
Finn, Ed, 103, 104, 125
Fontenelle, Bernard de, 159n23, 160n42
Form-Numbers, 24, 26–27, 28, 155n27
foundationalism, 160n33, 162n71
Fourier transform, 100
Fowler, David H., 13
Frege, Gottlob, 38, 39
functions
    visualization of, 46

Gaonkar, Dilip, 7, 158n8
Gauss, Carl Friedrich, 77, 83, 91, 96
    Disquisitiones arithmeticae, 96
geometric diagrams, 16

geometric proof, 19–20, 21, 22
geometry
    algebra and, 93
    apodeictic propositions, 30
    democracy and, 32
    nonarithmetic, 15
    Platonic view of, 14–15, 20, 29, 154n25
    as relations between spaces and forms, 14
Ghosh, Indradeep, 164n4
    "Banking on the Present," 166n33
Gillespie, Tarleton, 165n22
Gödel, Kurt, 12
Google, 130
Graham, S. Scott, 141
gravity, 147, 170n30
Greenspan, Alan, 111
Gries, Laurie, 141
Gross, Alan, 7, 169n15

Hacking, Ian, 12
Hanan, Joshua S., 164n4
    "Banking on the Present," 166n33
Haraway, Donna, 102
He, Jiankui, 2
Heidegger, Martin, 62, 82
Hersh, Reuben, 6, 36, 63–64, 65, 82, 162n71
Hiero II, Tyrant of Syracuse, 47–48
History of Mathematics, A (Boyer and Merzbach), 12
horizon of judgment, 108–9, 110, 144, 164n4
Horky, Phillip Sidney, 28
human experience
    mathematization of, 47
humans
    impact of cultural practices on, 147
    judgment of, 164n6
    nature of thought of, 44
    vs. nonhumans, 145
    as über-entities, 146, 150
    unique capacities of, 144–45, 146, 150
    as vital matter, 147

i (unit imaginary number)
    definition of, 5, 85
    Euler's equality, 91–92
imaginary numbers
    Descartes's view of, 91
    emergence of, 8, 10, 85–86, 101, 132–33, 138, 146
    importance of, 102, 163n29
    in mathematical network, 90, 91, 100–101
    as real, 95–96
    recurrence for, 153n36

imaginary numbers *(continued)*
  in science and technology, 98–101, 146, 163n33, 170n29
  solution of algebraic problems with, 89–90
  translative force of, 96, 101
incommensurability
  concept of, 142–43
incompleteness theorems, 12
infinitesimal quantity, 67, 69
infinitesimals
  as abstraction, 82
  Calculus and, 66–69, 161n47
  constitutive rhetoric of, 68, 80
  criticisms of, 68–69, 76, 160n33, 160n42
  emergence of, 138
  Euclidean geometry and, 64–65
  first use of, 60–61
  influence of, 67, 74, 79, 80, 81
  introduction of, 63, 64, 84
  mathematical meaning of, 82–83, 158n2, 159n16
  metaphors of, 68
  nonempirical status of, 82
  perception of, 78
  problem of definition of, 67–68, 69, 76, 77, 79
  ratios of, 71
  rhetorical features of, 60, 64, 65, 66, 74, 77–82, 83, 159n13
  scientific use of, 84
  theology and, 80–81
infinitesimals in degree, 68
infinity, 47
integral, 61

Jesseph, Douglas M., 160n33
Jones, Sam, 111
Jurek, Jakub, 166n36

Kahneman, Daniel, 126, 168n72
Kennedy, George, 6
Kennedy, J. B., 22, 23
Kline, Morris, 160n33, 160n41
knowledge
  as form of recollection, 15
  knowable world and, 138, 139
Kronecker, Leopold, 12
Krugman, Paul, 123

Lakatos, Imre, 5, 156n16
  Proofs and Refutations, 5, 6, 163n19, 171n37
Lakoff, George, 43, 44, 45, 46, 93–94, 132, 156n30
Lapérouse, Jean François de Galaup, comte de, 51

Latour, Bruno, 47
  on logos of geometric equality, 34
  on mathematics, 48, 53, 55, 132
  on modernism and understandings of science, 155n4
  on modernist metaphysics, 157n39
  notion of vincula, 151n12
  on numbers, 50
law of continuity, 69, 75
leading principle, 40
learning algorithms, 106
Leibniz, Gottfried Wilhelm
  Acta eruditorum, 60, 75
  on complex numbers, 91
  concept of Calculus, 60–62, 78, 81, 158n1
  criticism of, 65, 66, 68–69, 74, 76, 77, 79–80
  idea of infinitesimal, 61, 63, 64, 67, 74–78, 158n2, 160n41
  philosophical principle of, 76–77
  use of imaginary numbers, 85, 91
Lenglet, Marc, 108
Lewis, Michael
  The Big Short, 123
Li, David
  mathematical discourse of, 117, 125, 167n45
  "On Default Correlation," 104, 111, 113, 114, 117, 119
Li Gaussian copula
  acceptance of, 111, 112, 120, 166n36, 168n75
  agential force of, 126, 127–28, 168n75
  anchoring effect of, 126
  CDO market and, 167n56
  cognitive and cultural consequences of, 125–26
  discovery of, 125
  funnel-like structure of, 126
  horizon of judgment within, 105, 110, 115, 118, 128, 133
  loans and, 167n60
  materiality of, 120
  rhetorical approach to, 104
  rigor of, 125
  subprime mortgages and, 104, 123
  terraforming power of, 120, 122
limits
  mathematics of, 83
loans
  default correlation between, 118, 121
  pooling and tranching of, 121
logbooks, 51–52
logic of mimesis, 28
logic of representation, 86, 103

logos of geometric equality, 34
Lyne, John, 169n13, 169n20

mathematical black boxes, 151n10
mathematical concepts, 6
mathematical foundationalism, 68, 134
mathematical heuristic, 16
mathematical idea analysis, 43–44
mathematical innovation, 42, 134–35
mathematical mythos, 12–13
mathematical objects, 27, 38, 54, 55
mathematical persuasion, 40, 42–43
mathematical practice
   vs. mathematical forms, 34
   semiotic theory of, 42–43
mathematical proof, 20, 79, 161n66
mathematical propositions, 49, 50
mathematical realism
   characteristics of, 38–39
   influence of, 35, 134, 153n35, 157n33, 157n36
   limitations of, 5–6, 35
mathematical statements, 46, 54, 55–56
mathematical substructure, 32
mathematical symbols, 8, 38, 55
mathematical thinking, 39–40, 82
mathematics
   absolute truth and, 3, 108–9, 134, 136, 161n68
   absolute view of, 35
   as abstract thought, 30, 48
   agency in, 8–9, 153n33, 159n19
   articulation of recurrence, 99–100
   arts of justification in, 30
   calculation and deduction, 20
   capitalism and, 170n25
   cognitive-metaphorical approach to, 43–47
   computers and, 156n20
   constitutive approach to, 53, 157n47
   cultural practices and, 127, 171n40
   debate over fallibility of, 154n23
   definitions of, 36, 56
   education in, 163n29
   Euclidean *vs.* non-Euclidean, 84
   evolution of, 3–4, 10, 60, 78, 84–85, 94, 99, 133–36, 146, 159n19
   foundationalism in, 162n71
   function of, 149
   influence of, 83, 130, 153n35
   informal and formal discourse of, 156n16
   as innovation, 134–35
   inscriptive practices of, 4, 29, 50, 135, 139
   invention in, 142
   logic of representation and, 30, 54, 85, 86, 95, 133
   ontological force of, 106, 135, 137
   pedagogical approach to, 134
   Platonic view of, 11–13, 27–28, 29, 41, 44, 156n20
   politics and, 37, 48–49
   practice of, 49, 58
   productivity of, 52
   public culture and, 6, 9, 10, 53, 130–31, 132, 147–48
   Pythagorean principles of, 22–23
   reality-expanding power of, 86, 95–96, 124, 146
   reductionism of, 52–53
   relational thinking in, 136
   rhetoric and, 3–4, 5–7, 10–12, 32–33, 36–37, 55, 56, 59, 86, 101, 130–31, 133, 170n23
   rigor in, 160n33
   semiotic approach to, 39, 41–43
   static, 61
   studies of, 131–32
   symbolic action in dissemination of, 131–32
   theories of, 5
   translative rhetorical force of, 4–6, 9–10, 35, 47–53, 56–59, 131, 132–33, 135–36, 146, 149–50
   translative *vs.* realist understanding of, 49
   as vital matter, 147
Maxwell's equations, 99
mediation, 169n11
Merzbach, Uta, 12, 91, 96, 97
meta-Code, 42, 43
method of exhaustion, 16–17, 20, 22, 158n3
micropractices, 169n17
Miller, Carolyn, 141, 143
mnemotechnology, 105–6
Moore's Law, 141
mortgage-backed securities (MBSs), 112–13, 114, 121, 122, 123, 166n31, 167n60
mortgage defaults
   commensurable probability functions of, 117, 118
   human life analogy for, 114–15, 116, 117
   rates of, 115
   risk prediction, 117, 118
   scalability of, 115
   survival probability of, 118–19
motion
   vs. action, 153n34
mythos of mathematics, 12–13, 34, 42, 131

nature
vs. culture, 148, 152n15
navigation, 51–52, 53
negative numbers, 86–87, 88, 89, 94, 163n21
square root of, 86, 89, 90, 91
negative radicals, 91
Netz, Reviel, 29
Newton, Isaac
ambiguous language of, 73–74
concept of Calculus, 53, 60–62, 70–71, 73, 78, 81, 158n1
criticism of, 65, 66, 68–69, 70, 72, 73, 79–80, 161n44
on finite particles, 70
idea of infinitesimals, 61, 63, 64, 67, 70–74, 160n29, 160n41, 161n44
method of fluxions, 161n67
Methods of Fluxions and Infinite Series, 60
Principia, 70, 71, 72, 73, 80, 160n29, 161n67
Quadratura curvarum, 70
scientific views of, 69–70, 160n31
on ultimate velocity, 71, 72
use of imaginary numbers, 85, 91
Nieuwentijdt, Bernard, 66
numbers
anchoring effects of, 126, 168n72
characteristics of, 50–51
ideational meaning of, 22
imaginary decomposition of, 91
music and, 23
nonstandard, 83
odd and even, 25, 27
Platonic view of, 24–25
Pythagorean view of, 22, 23, 24
real vs. imaginary, 89
See also complex numbers; negative numbers
Núñez, Rafael, 43, 44, 45, 46, 93–94, 132, 156n30
Nussbaum, Martha, 31, 157n34

objective thought, 38–39
"On Default Correlation" (Li), 111, 113, 114, 117, 119
O'Neil, Cathy, 107
ontological hierarchy, 145

Pascal, Blaise, 75
Pasquale, Frank, 106, 165n14
Peirce, Charles Sanders, 39
Philolaus, 23
phronesis (practical judgment), 105, 118

π (mathematical constant)
approximation of, 17
definition of, 92, 93–94
as expression of unit circle, 92–93
as measure of periodicity, 93–94
as numerical signifier of angles, 93
Plato
on arts of innovation and justification, 30
on calculation, 11
Cratylus, 28
dialectic of, 18, 26, 27
Epinomis, 30
epistemology of, 154n25
evolution of views of, 154n24
filtering strategy, 29
geometric logos of, 14–16, 17–18, 20, 27, 30, 31–32, 34, 154n25
Gorgias (Plato), 11, 28, 31–32, 33, 34
on images, 28, 155n28
influence of, 35
Meno, 11, 13, 15–16, 17–18, 21–22, 23, 27, 28, 29, 154n15
metaphysics of, 10, 27
on nature of knowledge, 26
on numbers, 24–25, 26–27
on number three, 155n26
Phaedo, 11, 27
Phaedrus, 28
philosophy of mathematics, 11–13, 26, 27–28, 30, 34, 131, 133, 154n9, 154n15, 156n20
proof of immortality of soul, 11, 18, 23–24, 27
Pythagoreanism and, 22
Republic, 11, 14, 25, 26, 27–28, 154n15
on rhetoric and mathematics, 32–33, 59
science of measurement, 31
structure of dialogues of, 22, 23
Theaetetus, 11, 18
theory of Forms, 11, 18, 24, 26–27, 28, 29–30, 31, 154n14
theory of language, 29, 59
Timaeus, 11, 29
treatment of geometry, 13–14
on words and symbols, 28, 29–30
Platonic Academy in Athens, 12
Platonic realism, 37, 38, 41, 44–45, 59, 60, 67
Platonism
forms of, 153n4
polyhedra, 171n37
pooling, 121

potential relations, 51
principle of composition, 58, 144, 151n12
proof, 40–41, 42
    See also mathematical proof
pure mathematics, 3, 133
Pythagoreanism, 22–23, 24

quantum physics, 100

ratios, 50, 64–65, 159n27
ratios of finite differences, 71
ratios of infinitesimals, 71, 72, 77, 79
realism
    application to economics, 166n36
    vs. relativism, 140
reality
    as network of relations, 146–47, 148
    plasticity of, 169n13
    social construction of, 145
Reid, Rob, 106
Reimann mapping theorem, 100
remembrance, 105–6
representation
    dominance of, 81
    episteme of, 139
    limits of, 81, 86, 99
representationalism, 138, 157n47, 169n10
rhetoric
    anthropocentric theories of, 7–8
    constitutive, 53, 54, 57, 58, 132, 159n16
    definition of, 36, 143
    as embodied inscriptive practice, 135, 138, 139–41
    emergence of, 145
    eristic virtue of, 33
    evolution of, 140–41, 143–44
    as false art, 33
    limitations of, 145
    mathematics and, 3–4, 5–7, 10–12, 32–33, 36–37, 55, 56, 59, 86, 101, 130–31, 133, 170n23
    as multiplicity, 157n47, 171n38
    nonhuman entities and, 141
    purpose of, 143
    representative, 54, 145
    in science and technology, 143
    semiotic element of, 39
    speech and, 7–8, 138
    studies of, 6–8, 148–49, 158n8
    theory of, 4, 43, 131, 132, 141, 153n34
    translative vs. traditional, 141, 142
rhetoric in mathematics, 9

rhetoric of mathematics, 9, 63–64
Rice, Jenny, 141
Rickert, Thomas, 141
Rivers, Nathaniel, 141
Robinson, Abraham, 83
Rorty, Richard, 83
Rotman, Brian
    books of, 39
    on mathematical proof, 40–41
    on science of mathematics, 12, 37, 38, 42, 43
Russell, Bertrand, 36, 157n33
Russell, Stuart, 106

Salmon, Felix, 110, 166n36
Schiappa, Edward, 9
science of judgment, 155n34
self-interest, 171n42
semiotics in mathematics, 39, 40, 41
Series, Caroline, 163n33, 170n29
sine-wave function, 92–93, 93
Singh, Simon, 85
situational rhetoric, 65
social constructionism, 143, 170n20
Socrates
    method of question and answer, 21–22
    proof of everlasting soul, 23–24
Sophists, 32, 33
speech
    as human capacity, 145
    in rhetoric, 135, 138
square root of negative numbers, 86, 89, 90, 91
Stafford, Erik, 166n36
Stewart, Ian, 80, 160n41
Stiegler, Bernard, 105
Stormer, Nathan, 127, 157n47
Strogatz, Steven, 91
Subject
    relationship between Agent and, 39, 40, 41
subprime mortgages, 123
symbolic action, 58, 127, 138, 171n37
systems
    scalable vs. nonscalable, 124

Taleb, Nassim, 124
technē
    forces of, 30–31
Thomas, Neal, 105
thought
    thinker and, 38–39
Thurston, William, 6, 171n37
tranching technique, 111–12, 121

translative rhetoric
　development of, 137–38
　diagrammatic model of, *57, 137*
　diagram of, *128*
　mathematics as, 56–57, 58, 132
　notion of, 4, 10, 140
　rhetorical inquiry and, 143–44
　theory of, 141, 170n21
trigonometry, 91, 92–93, 163n18
true knowledge, 18
truth, absolute
　as purpose of mathematics, 3, 108–9, 134, 136, 161n68
tuchē
　forces of, 30–31
Tversky, Amos, 126, 168n72

ultimate velocity, 71, 72
unit circle
　definition of, 92
　Euler's identity and, 94
　π and, 92, 92–93
　translation into sine function, 92–93, *93*

velocity, 61, 71, 72
Vilenkin, Naum, 81
vinculum, 5, 9, 50, 86, 151n7, 151n12
virtues, 33
vitality, 147
Vlastos, Gregory, 22, 154n14, 154n25

Wallis, John, 95
　Treatise of Algebra, 163n21
Walsh, Lynda, 141
Watson, James, 99
Wessel, Caspar, 96
Where Mathematics Comes From (Lakoff and Núñez), 43–47
Wigner, Eugene, 42
Williams, Robert Julian Borchak, 1

www.ingramcontent.com/pod-product-compliance
Lightning Source LLC
Chambersburg PA
CBHW022057290426
44109CB00014B/1135